S. Faedo (Ed.)

# il principio di minimo e sue applicazioni alle equazioni funzionali

Lectures given at the
Centro Internazionale Matematico Estivo (C.I.M.E.),
held in Pisa, Italy,
September 1-10, 1958

Springer

FONDAZIONE
CIME
ROBERTO CONTI

C.I.M.E. Foundation
c/o Dipartimento di Matematica "U. Dini"
Viale Morgagni n. 67/a
50134 Firenze
Italy
**cime@math.unifi.it**

ISBN 978-3-642-10924-9 e-ISBN: 978-3-642-10926-3
DOI:10.1007/978-3-642-10926-3
Springer Heidelberg Dordrecht London New York


Reprint of the 1st ed. C.I.M.E., Ed. Cremonese, Roma, 1958.
With kind permission of C.I.M.E.

Printed on acid-free paper

Springer.com

CENTRO INTERNATIONALE MATEMATICO ESTIVO
(C.I.M.E)

Reprint of the 1st ed.- Pisa, Italy, September 1-10, 1958

# IL PRINCIPIO DI MINIMO E SUE APPLICAZIONI ALLE EQUAZIONI FUNZIONALI

Estratto dagli *Annali della Scuola Normale Superiore di Pisa*
Serie III. Vol. XIII. Fasc. II (1959)

# ON ELLIPTIC PARTIAL DIFFERENTIAL EQUATIONS

by L. NIRENBERG (New York) (*)

## Outline.

This series of lectures will touch on a number of topics in the theory of elliptic differential equations. In Lecture I we discuss the fundamental solution for equations with constant coefficients. Lecture 2 is concerned with Calculus inequalities including the well known ones of Sobolev. In lectures 3 and 4 we present the Hilbert space approach to the Dirichlet problem for strongly elliptic systems, and describe various inequalities. Lectures 5 and 6 comprise a self contained proof of the well known fact that « weak » solutions of elliptic equations with sufficiently « smooth » coefficients are classical solutions.

In Lectures 7 and 8 we describe some work of Agmon, Douglis, Nirenberg [14] concerning estimates near the boundary for solutions of elliptic equations satisfying boundary conditions. This work is based on explicit formulas, given by Poisson kernels, for solutions of homogeneous elliptic equation with constant coefficients in a half space.

Throughout, for simplicity we treat one equation in one unknown. The material will on the whole be self contained, though of course not all proofs can be included. However, we shall attempt to indicate those of the main results.

---

(*) Questo ciclo di conferenze è stato tenuto a Pisa dal 1º al 10 settembre 1958, e ha fatto parte del corso del C. I. M. E. che ha avuto per tema: « Il principio di minimo e sue applicazioni alle equazioni funzionali ». Tale corso si è svolto in collaborazione con la Scuola Narmale Superiore e l'Istituto Matematico dell'Università di Pisa. In questi Annali saranno successivamente pubblicati i corsi di conferenze tenuti dai professori C. B. Morrey e L. Bers.

## Lecture I. The Fundamental Solution.

I would like to start with a few general and somewhat unrelated comments. In studying differential equations one is usually interested in obtaining *unique* solutions by imposing suitable boundary or initial conditions, the kind depending on the so-called «type» of the equation - elliptic, hyperbolic, etc. However, the type classification for general equations has not been carried out, and in many cases it is not known what boundary conditions to impose. Indeed for equations that change type — and we are all familliar with the initial work in this field due to Professor Tricomi — the nature of the boundary conditions is far from obvious.

Thus if one considers an arbitrary equation without regard to type it is a natural question to ask whether there exist solutions at all. In fact there are occasions when one simply wants some solutions. Such occur often in differential geometry. Take a well known case: to introduce isothermal coordinates with respect to a given Riemannean metric on a two dimensional manifold. This reduces to a local problem of finding nontrivial solutions of a differential equation in a neighborhood of a point.

Another question is: are there solutions in the large of a given equation. For the preceding this is answered by uniformization theory for Riemann surfaces.

In this talk we will consider for some special cases the question: For a given differential operator $L$ are there solutions of $Lu = f$ for « well behaved » functions $f$. Of course equations with analytic coefficients always have local solutions, obtained for instance by power series expansions (Cauchy-Kowalewski).

Recently Hans Lewy [1] exhibited an equation with $C^\infty$ coefficients having no solutions ewen locally. Since it is easy to describe, we present it:

In 3-space with coordinates $x, y, t$, set $z = x + iy$, write the Cauchy-Riemann operator as $\dfrac{\partial}{\partial \bar z} = \dfrac{1}{2}\left(\dfrac{\partial}{\partial x} + i\dfrac{\partial}{\partial y}\right)$, and consider the differential equation

$$Lu = \left(\frac{\partial}{\partial \bar z} + i z \frac{\partial}{\partial t}\right) u = \frac{\partial \psi(t)}{\partial t}$$

where the right hand side is a continuous real function of $t$ alone which, for convenience, is written as a derivative of a real function $\psi$.

THEOREM: *If there is a continuously differentiable solution $u$ of the equation in a neighborhood of the origin, then $\psi(t)$ is real analytic.*

Thus for any non-analytic $\psi$ there is no solution near the origin. (The proof may be easily modified to show that there are also no « generalized solutions »).

*Proof*: If we integrate $\dfrac{\partial u}{\partial \bar{z}} d\theta$ over a circle $|z|^2 = s \geq 0$, $z = s^{1/2} e^{i\theta}$, we establish easily the identity

$$\int_0^{2\pi} \frac{\partial u}{\partial \bar{z}} d\theta = \frac{\partial}{\partial s} \int_0^{2\pi} z\, u\, d\theta .$$

Now set $\zeta = s + i\,t$ and $U(\zeta) = \int z\, u\, d\theta$. Integrating the equation for $u$ over the circle we find that $U$ satisfies

$$\left(\frac{\partial}{\partial s} + i \frac{\partial}{\partial t}\right) U = 2\pi \frac{d\psi}{dt}$$

or

$$\left(\frac{\partial}{\partial s} + i \frac{\partial}{\partial t}\right)(U + 2\,\pi\, i\, \psi) = 0 .$$

It follows that $V(\zeta) = U + 2\,\pi\, i\, \psi$ is a holomorphic function of $\zeta = s + i\,t$ in a domain near the origin with re $\zeta = s > 0$. But on $s = 0$ the function $U$, i. e. the real part of $V$, vanishes, and therefore $V$ can be continued analytically across $s = 0$. Hence $\psi$ is analytic.

In [1] Lewy also constructs a function $F$ such that the equation $Lu = F$ has no « smooth » solution in the neighborhood of any point. Lewy also conjectures that there are *homogeneous* equations with $C^\infty$ coefficients having no solutions in the neighborhood of any point.

The simplest class of differential operators $L$ of arbitrary type, for which one might expect solutions $u$ of

$$Lu = f \tag{1.1}$$

to exist, for all well behaved functions $f$, are operators with constant coefficients. In the last few years a considerable study has been made of general differential operators with constant coefficients. (See Ehrenpreis [2], Hörmander [3], Malgrange [4]. Solutions of (1.1) can be found, at least locally, if one knows that a fundamental solution $E$ of $L\,E = \delta$ (the Dirac $\delta$ function) exists. This is a (possibly generalized) function $E$ such that

$$E * Lu = u$$

for all $C^\infty$ functions $u$ with compact support. We shall denote the class of such functions by $C_0^\infty$. Here $*$ denotes convolution. Then if $f$ is in $C_0^\infty$ the function $u = E * f$ is a solution of (1.1).

Malgrange [4] and Ehrenpreis [2] proved the existence of a fundamental solution for any differential operator with constant coefficients. However it is not difficult to construct one explicitily, as Hörmander, and also Trêves [5], have shown, and we shall now describe such a construction.

First we fix our

Notation: We consider functions $u(x)$ of $n$ variables $x = (x_1, \dots, x_n)$ and denote the differentiation vector by $D = (D_1, \dots, D_n)$, $D_i = \partial/\partial x_i$. The letters $\beta, \gamma, \mu, \nu$ will denote vectors $\beta = (\beta_1, \dots, \beta_n)$ with non-negative integral coefficients $\beta_i$, and we set $|\beta| = \Sigma \beta_i$. Otherwise for any vector $\xi = (\xi_1, \dots, \xi_n)$, $|\xi|$ will represent its Euclidean length $|\xi|^2 = \Sigma |\xi_i|^2$, and $\xi \cdot \eta = \Sigma \xi_i \eta_i$. We write

$$\xi^\beta = \xi_1^{\beta_1} \dots \xi_n^{\beta_n}, \qquad D^\beta = D_1^{\beta_1} \dots D_n^{\beta_n};$$

for convenience we shall also, on occasion, express a general $m^{th}$ order partial derivative of a function $u$ by $D^m u \cdot C_0^\infty$ will denote the class of $C^\infty$ functions with compact support.

We consider now a differential operator $L$ of order $k$ with constant coefficients, which we may write as a polynomial in $D$ of order $k$.

$$L = L(D).$$

In constructing the fundamental solution let us first argue in a heuristic manner. Introduce the Fourier transform of the function $u(x)$

$$\tilde{u}(\xi) = \int e^{-ix\cdot\xi} u(x)\, dx,$$

integration being over the entire $n$-space. Then

$$\widetilde{L(D)u} = L(i\xi)\, \tilde{u}(\xi).$$

So if $u = E * Lu = \int E(x-y)\, Lu(y)\, dy$ then

$$\tilde{u} = \tilde{E}(\xi)\, L(i\xi)\, \tilde{u}(\xi)$$

or

$$\tilde{E} = \frac{1}{L(i\xi)},$$

or

$$E(x) = (2\pi)^{-n} \int \frac{e^{ix\cdot\xi}}{L(i\xi)}\, d\xi\,. \tag{1.2}$$

*Problem : give formula* (1.2) *a meaning.*

In attempting to do this (and there are many ways) there are two difficulties that occur. The first is the non-integrability at infinity, duo to the fact that we are integrating over the full $n$-space. The second difficulty is caused by the real roots $\xi$ of the polynomial $L(i\xi)$.

The first difficulty is easily overcome. It essentially expresses the fact that is general $E$ is a distribution, i. e. a finite derivative of a continuous function. Instead of constructing $E$ directly we shall construct the fundamental solution $E_N$ of the operator $(1-\Delta)^N L = (1 - \sum_i D_i^2)^N L(D)$. We shall construct a fundamental solution $E_N$ having continuous derivatives up to any given order, by taking $N$ sufficiently large. We may then take, in the distribution sense,

$$E = (1-\Delta)^N E_N\,, \tag{1.3}$$

i. e. for $f$ in $C_0^\infty$ the function

$$u = E_N * (1-\Delta)^N f$$

is a solution of $Lu = f$.

Thus we consider, for $p(\xi) = 1 + \sum \xi_j^2$

$$E_N = (2\pi)^{-n} \int \frac{e^{ix\cdot\xi}}{p^N(\xi)\, L(i\xi)}\, d\xi\,. \tag{1.4}$$

Taking $N$ large eliminates the first difficulty, i. e. the trouble at infinity.

Now to handle the second difficulty. We may assume, after a possible rotation of coordinates, that the coefficient of $D_n^k$ in $L(D)$ is $\neq 0$, say unity. Consider $L(i\xi)$ as a polynomial in $\xi_n$. We shall first integrate in (1.4) with respect to the variable $\xi_n$, keeping $\xi' = (\xi_1, \dots, \xi_{n-1})$ fixed, however we shall move the line of integration from the real line to a parallel line lying in the complex $\xi_n$ plane.

For fixed real $\xi'$ there are $k$ roots $\xi_n$ of $L(i\xi)$. In the strip $|\mathfrak{Im}\,\xi_n| \leq \frac{1}{2}$ in the complex $\xi_n$ plane there is therefore a line parallel to the real axis whose distance from any root is at least $(2k+2)^{-1}$, as one easily sees. Let us a choose one such line $\mathfrak{Im}\,\xi_n = c(\xi')$ whose distance to any root is at least $(4k+4)^{-1}$. The choice of $c(\xi')$ depends on $\xi'$, but it is easy to see that $c = c(\xi')$ may be chosen so as to be continuous except on a set of $\xi'$ of $(n-1)$-dimensional measure zero.

Setting $\eta = \eta(\xi') = (0, \dots, c(\xi'))$ we now take as definition

$$(1.4)' \qquad E_N = (2\pi)^{-n} \int \frac{e^{ix\cdot(\xi + i\eta(\xi'))}}{p^N(\xi + i\eta)\, L(i(\xi + i\eta))}\, d\xi$$

where integration is first with respect to $\xi_n$.

Since

$$|p(\xi + i\eta(\xi'))| \geq \frac{3}{4} \quad \text{and} \quad |L(i(\xi + i\eta))| \geq (4k+4)^{-k}$$

we see that $E_N$ has derivatives up to any given order, if $N$ is large enough.

We have finally to verify that for $u \in C_0^\infty$

$$u = E_N * (1-\Delta)^N Lu \equiv \int E_N(x-y)(1-\Delta)^N Lu(y)\, dy.$$

Setting $(1-\Delta)^N L(D) = L_N(D)$, the right hand side equals

$$(2\pi)^{-n} \iint \frac{e^{i(x-y)\cdot(\xi+i\eta)}}{L_N(i(\xi+i\eta))}\, d\xi\, L_N(D)\, u(y)\, dy.$$

Since $u$ has compact support its Fourier transform $\tilde{u}(\xi)$ can be extended to complex vectors $\xi$ as an entire analytic function, and since $u \in C^\infty$ the derivatives of $\tilde{u}$ die down faster that any power of $|\xi|$ as we go to infinity in a strip $|\mathfrak{Im}\, \xi| <$ constant. Thus, interchanging the order of integration in the above, we find that it equals

$$(2\pi)^{-n} \int \frac{e^{ix(\xi+i\eta)}}{L_N(i(\xi+i\eta))} L_N(i(\xi+i\eta))\, \tilde{u}(\xi+i\eta)\, d\xi =$$

$$= (2\pi)^{-n} \int e^{ix(\xi+i\eta)}\, \tilde{u}(\xi+i\eta)\, d\xi.$$

Because of the behaviour of $\tilde{u}$ of infinity we may shift the line of integration of the $\xi_n$ parallel to itself and find that this expression

$$= (2\pi)^{-n} \int e^{ix\cdot\xi}\, \tilde{u}(\xi)\, d\xi = u(x).$$

Thus the function $E_N$ defined by (1.4)′ is a fundamental solution for the operator $L_N$. The desired fundamental solution of $Lu$ then is given by (1.3).

One sees easily that the fundamental solution $E_N$ given by (1.4)′ has exponential growth in the $x_n$ variable.

For further important work on fundamental solutions for equations with constant coefficients we refer to Hörmander [6].

Consider now elliptic differential operators with constant coefficients. These are operators $L$ whose leading part $L'$ — consisting of the terms of highest order — satisfy

$$L'(\xi) \neq 0 \quad \text{for real} \quad \xi \neq 0 .$$

We shall have need later of the fundamental solution for a homogeneous elliptic operator with constant coefficients, i. e. $L' = L$. For such, of course, the fundamental solution first constructed by Herglotz is well behaved at infinity. We shall use the following form of it, given in F. John's book [7].

$$E(x) = -\frac{1}{(2\pi i)^n (k+q)!} \Delta^{\frac{n+q}{2}} \int\limits_{|\xi|=1} \frac{(x \cdot \xi)^{k+q}}{L(\xi)} \log \frac{x \cdot \xi}{i} \, d\,\omega_\xi \tag{1.5}$$

where integration is over the full unit sphere with $d\,\omega_\xi$ as the element of area, $q$ is a non-negative integer of the same parity $n$, i. e. $q+n$ is even, and the principal branch of the logarithm is taken with the plane slit along the negative real axis.

From (1.5) we obtain as a special case, for $L = \Delta$ power, the following identity which is due to $F$. John and used extensively in [7], representing the $\delta$ function in terms of plane waves: For $u$ in $C_0^\infty$

$$u = -\frac{1}{(2\pi i)^n\, q!} \Delta^{\frac{n+q}{2}} \left[ \int\limits_{|\xi|=1} (x \cdot \xi)^q \log \frac{x \cdot \xi}{i} \, d\,\omega_\xi * u \right] . \tag{1.6}$$

In [7] John derives (1.6) from the known expression for the fundamental solution for a power of the Laplacean, and then derives (1.5) from (1.6). This may be done as follows. Suppose $K(x \cdot \xi)$ satisfies

$$L\,K(x \cdot \xi) = (x \cdot \xi)^q \log \frac{x \cdot \xi}{i} ,$$

then a fundamental solution of the operator $L$ is given by

$$-\frac{1}{(2\pi i)^n\, q!} \Delta^{\frac{n+q}{2}} \int\limits_{|\xi|=1} K(x \cdot \xi) \, d\,\omega_\xi .$$

But such a $K$ is easily found. If we set $x \cdot \xi = \sigma$ then $K(\sigma)$ satisfies

$$L(\xi)\left(\frac{d}{d\sigma}\right)^k K(\sigma) = \sigma^q \log \sigma/i\,,$$

a solution of which is

$$K(\sigma) = \frac{1}{L(\xi)} \frac{q\,!}{(k+q)!}\,\sigma^{k+q}\left(\log \frac{\sigma}{i} + c_{k,q}\right),$$

with $c_{k,q}$ an appropriate constant. If we insert this into the above expression for the fundamental solution of $L$ we obtain the expression

$$-\frac{1}{(2\pi i)^n (k+q)\,!}\,\Delta^{\frac{n+q}{2}} \int\limits_{|\xi|=1} \frac{(x \cdot \xi)^{k+q}}{L(\xi)}\left(\log \frac{x \cdot \xi}{i} + c_{k,q}\right)$$

which differs from (1.5) only by the term involving $c_{k,q}$. But this term is a polynomial of degree $k - n$ which is therefore a solution of $Lv = 0$, and so may be ignored.

It should also be possible to derive (1.5) from the heuristic formula (1.2). (1.5) aserts that

$$-\frac{1}{(2\pi i)^n (k+q)\,!} \int\limits_{|\xi|=1} \frac{(x \cdot \xi)^{k+q}}{L(\xi)} \log \frac{x \cdot \xi}{i}\, d\,\omega_\xi \tag{1.7}$$

is a fundamental solution for the operator $\Delta^{\frac{n+q}{2}} L$. Let us attempt to derive this expression from the corresponding expression of (1.2):

$$(-1)^{\frac{n+q}{2}} (2\pi)^{-n} \int \frac{e^{ix\cdot\xi}}{|\,\xi\,|^{n+q} L(i\,\xi)}\, d\,\xi\,. \tag{1.8}$$

Arguing heurisitically again let us modify the expression by introducing polar coordinates in the $\xi$ space

$$\xi = \varrho\,\eta\,, \qquad \varrho = |\,\xi\,|\,, \qquad |\,\eta\,| = 1\,.$$

Then (1.8) becomes

$$(-1)^{n+q+k} (2\pi)^{-n} \int\limits_{|\eta|=1} \int\limits_0^\infty \frac{e^{i\varrho x\cdot\eta}}{L(\eta)}\, \varrho^{-1-q-k}\, d\,\varrho\, d\,\omega_\eta\,. \tag{1.8'}$$

Let us now write the heuristic expression

$$\int_0^\infty e^{i\varrho x\cdot\eta}\,\varrho^{-1-q-k}\,d\,\varrho \tag{1.9}$$

as a well defined contour integral

$$\frac{1}{2\pi i}\int_{\mathfrak{C}} e^{i\varrho x\cdot\eta}\,\varrho^{-1-q-k}\,(\log(-\varrho)+c) \tag{1.9$'$}$$

where the contour $\mathfrak{C}$ is a curve which goes from $+\infty$ in the complex $\varrho$ plane, encircles the origin counterclockwise and returns to $+\infty$ along the real axis, the branch for the logarithm is the same as above, and the constant $c$ is chosen so that

$$\int_{\mathfrak{C}} e^{i\varrho}\,\varrho^{-1-q-k}\left(\log(-\varrho)+c-\frac{i\pi}{2}\right)d\varrho=0\,.$$

The expressions (1.9)′ may be evaluated explicity, and on insertion into (1.8)′, yields the expression (1.7). We leave the calculation to the reader.

## Lecture II. Calculus Inequalities.

A priori estimates play a central role in the theory of partial differential equations. They are of various kinds — pointwise estimates for derivatives of solutions and their modulus of continuity, and estimates of, say, $L_p$ norms of solutions and their derivatives — and it is naturally important to understand the relationships between these various estimates.

For instance, the well known results of Sobolev assert that if the m'th order derivatives $D^m u$ of a function $u(x_1,\dots,x_n)$ (with compact support) are in $L_r$, $1<r<\infty$ then lower order derivatives $D^j u$, $j<m$ belong to $L_p$ for some $p$, or, if $r$ is sufficiently high, the $D^j u$ are bounded and satisfy a Hölder condition with a certain exponent $\alpha$.

Since we shall often make use of it, let us recall here the notion of

Hölder continuity. A function $f(x)$ defined on a set $S$ in a Euclidean space satisfies a Hölder condition there with exponent $\alpha$, $0<\alpha<1$, if

$$[f]_\alpha=[f]_\alpha^S=\underset{x,y\in S}{e\,.\,u\,.\,b}\,\frac{|f(x)-f(y)|}{|x-y|^\alpha} \tag{2.1}$$

is finite. It is Hölder continuous (exponent $\alpha$) in a domain if it satisfies a Hölder condition with exponent $\alpha$ in every cempact subset of the domain.

This lecture is concerned with calculus inequalities relating integral and pointwise estimates of functions and their derivatives. The recent important result of de Giorgi [11] on the differentiability of solutions of regular variational problems seems in fact to be based on a calculus inequality asserting that certain integral estimates imply Hölder continuity. We shall consider functions $u(x)$ defined in $n$-dimensional Euclidean space and belonging to $L_q$, and whose derivatives of order $m$ belong to $L_r$, $1 \leq q, r \leq \infty$. We shall present interpolative inequalities for the $L_p$ and Hölder norms $[\ ]_\alpha$ of derivatives $D^j u$, $0 \leq j < m$, for the maximal range of $p$ and $\alpha$. Our inequalities are a combination of, and include, those usually called of Sobolev type (which hold also for fractional derivatives, and rather straightforward proofs of which may be found in [8]), and familiar interpolative inequalities such as

$$M_1^2 \leq \text{constant } M_0 \cdot M_2$$

where $M_i$ is $e.u.b.$ of the $L_p$ norms of the derivatives of order $i$ of a function $u$, $i = 0, 1, 2$. The proofs use only first principles and are entirely elementaty. (No attempt will be made here to obtain best constants). The inequalities is this section were presented at the Int'e Congress in Edinburgh August 1958, where we learned that almost equivalent results had also been proved by E. Gagliardo.

In this lecture we shall use the following

Notation: For $-\infty < \dfrac{1}{p} < \infty$ we defins the norms and seminorms $|u|_p$ for functions $u(x)$ defined in a domain $\mathcal{D}$ in $n$-dimensional spaces:

For $p > 0$

$$|u|_p = \text{the } L_p \text{ norm of } u \text{ in } \mathcal{D}.$$

$$= \left( \int_{\mathcal{D}} |u|^p \, dx \right)^{\frac{1}{p}}.$$

For $p < 0$ set $s = [-n/p]$, $-\alpha = s + n/p$ and define

$$|u|_p = e.u.b.\, [D^s u]_\alpha^{\mathcal{D}} \qquad \text{if } \alpha > 0,$$

$$|u|_p = e.u.b.\, |D^s u| \qquad \text{if } \alpha = 0,$$

where $e.u.b.$ is taken with respect to all partial derivatives $D^s$ of order $s$, and over points in $\mathcal{D}$.

We define $|D^j u|_p$ as the maximum of the $|\ |_p$ norms of all $j$-th order derivatives of $u$.

We shall express our result for functions $u$ defined in the entire $n$-space $E^n$. Extension to other domains will be described briefly in the remarks after the theorem.

THEOREM: *Let $u$ belong to $L_q$ in $E^n$ and its derivatives of order $m$, $D^m u$, belong to $L_r$, $1 \leq q, r \leq \infty$. For the derivatives $D^j u$, $0 \leq j < m$, the following inequalities hold*

$$|D^j u|_p \leq \text{constant}\, |D^m u|_r^a\, |u|_q^{1-a}, \tag{2.2}$$

where

$$\frac{1}{p} = \frac{j}{n} + a\left(\frac{1}{r} - \frac{m}{n}\right) + (1-a)\frac{1}{q},$$

*for all $a$ in the interval*

$$\frac{j}{m} \leq a \leq 1 \tag{2.3}$$

*(the constant depending only on $n, m, j, q, r, a$), with the following exceptional cases*

1. *If $j = 0$, $rm < n$, $q = \infty$ then we make the additional assumption that either $u$ tends to zero at infinity or $u \in L_{\tilde{q}}$ for some finite $\tilde{q} > 0$.*

2. *If $1 < r < \infty$, and $m - j - n/r$ is a non negative integer then* (2.2) *holds only for $a$ satisfying $j/m \leq a < 1$.*

We shall not give a complete proof of the theorem here but shall indicate the main steps. First some comments.

1. The value of $p$ is determined simply by dimensional analysis.

2. For $a = 1$ the fact that $u$ is contained in $L_q$ does not enter in the estimate (2.2), and the estimate is equivalent to the results of Sobolev (note that we permit $r$ to be unity).

3. That $j/m$ is the smallest possible value for $a$ may be seen by taking $u = \sin \lambda x_1 \zeta(x)$ where $\zeta$ is in $C_0^\infty$: For large $\lambda$ we have $|u|_q = 0(1)$, $|D^j u|_p = 0(\lambda^j)$, $|D^m u|_r = 0(\lambda^m)$ where no $0$ can be replaced by $o$.

4. It will be clear from the proof that the result holds also for $u$ defined in a product domain

$$-\infty < x_s < \infty,\ 0 < x_t < \infty\ :\ s = 1, \dots, k:\ t = k+1, \dots, n,$$

and hence for any domain that can be mapped in a one-to-one way onto such a domain by a sufficiently «nice» mapping.

5. For a bounded domain (with « smooth » boundary) the result holds if we add to the right side of (2.1) the term

$$\text{constant} \, | u |_{\tilde{q}} .$$

for any $\tilde{q} > 0$. The constants then depend also on the domain.

6. Similar estimates hold for the $L_p$ norms of $D^j u$ on linear subspaces of lower dimension, for suitable $p$.

7. Similar interpolation inequalities also hold for fractional derivatives, but their proof is not so elementary.

The theorem, in its full generality should be useful in treating nonlinear problems. We mention in particular that from (2.2) for $a = j/m$, $q = \infty$ it follows that the set of functions $u$ which are bounded and have derivatives of order $m$ belonging to $L_r$ forms a Banach Algebra. For $r = 2$ this is called the Schauder ring.

The proof of the theorem is elementary and contains in particular an elementary proof for the Sobolev case $a = 1$. In order to prove (2.2) for any given $j$ one has only to prove it for the extreme values of $a$, $j/m$ and unity. (If Case 2 holds some additional remark has to be made.) For in general there is a simple

*Interpolation Lemma: if* $-\infty < \lambda \leq \mu \leq \nu < \infty$ *then*

$$| u |_{\frac{1}{\mu}} \leq c \, | u |_{\frac{1}{\lambda}}^{\frac{\nu-\mu}{\nu-\lambda}} \cdot | u |_{\frac{1}{\nu}}^{\frac{\mu-\lambda}{\nu-\lambda}}$$

*where* $c$ *is independent of* $u$.

The lemma is easily proved; for $\lambda > 0$ it is merely the usual interpolation ineqnality for $L_p$ norms.

Let us turn now to the proof of the theorem, or at least to the main points. Consider first the Sobolev case, $a = 1$. It suffices to consider the case $j = 0$, $m = 1$, from which the general result may then be derived. If $r > n$ (2.2) asserts that $u$ satisfies a certain Hölder condition, and an elementary proof due to Morrey has long been known. We shall sketch it here for functions defined in a general domain $\mathcal{D}$.

*Definition:* A domain $\mathcal{D}$ is said to have the strong cone property if there exist positive constants $d$, $\lambda$ and a closed solid right spherical cone $V$ of fixed opening and height such that any points $P$, $Q$ in $\overline{\mathcal{D}}$ (the closure of $\mathcal{D}$) with

$$| P = Q | \leq d$$

are vertices of cones $V_P$, $V_Q$ lying in $\bar{\mathcal{D}}$ which are conrguent to $V$ and have the following property: the volume of the intersection of the sets: $V_P$, $V_Q$, and the two spheres with conters $P, Q$ and radius $|P-Q|$, is not less than $\lambda |P-Q|^n$.

We now prove the assertion

*If $u$ has first derivatives in $L_r$, $r > n$, in a domain $\mathcal{D}$ having the strong cone property, then for points $P, Q$ in $\mathcal{D}$ with $|P-Q| \leq d$, we have*

$$\frac{|u(P)-u(Q)|}{|P-Q|^{1-\frac{n}{r}}} \leq \text{constant} \, |Du|_r$$

*where the constant depends only on $d$, $\lambda$, $V$, $n$ and $r$.*

(From this follows easily an estimate for $[u]_{1-\frac{n}{r}}$, depending on the domain).

*Proof:* Set $s = |P-Q|$ and let $S_P (S_Q)$ be the intersection of $V_P (V_Q)$ with the sphere about $P(Q)$ radius $s$. Set $S_P \cap S_Q = S$. If $R$ is a point in $S$ we have, on integrating with respect to $R$ over $S$,

$$\text{Volume of } S \cdot |u(P)-u(Q)| \leq \int_S |u(P)-u(R)| \, dR +$$

$$+ \int_S |u(R)-u(Q)| \, dR.$$

Because of the strong cone property the left hand side is not less than

$$\lambda s^n |u(P)-u(Q)|.$$

The first term on the right may be estimated as follows. Introducing polar coordinates $\varrho, \eta$, about $P$, where $\eta$ is a unit vector, we find easily that the first term in the right is bounded by

$$\int_{S_P} \varrho^{n-1} \, d\omega_\eta \, d\varrho \int_0^\varrho \left| \frac{\partial u}{\partial \varrho} \right| d\varrho \leq \text{constant } s^n \int_{S_P} \left| \frac{\partial u}{\partial \varrho} \right| \frac{dx}{\varrho^{n-1}}$$

(where $d\omega$ is the element of area on the unit sphere, and $dx$ is the element of volume)

$$\leq \text{constant } s^n \left( \int_{S_P} \left| \frac{\partial u}{\partial \varrho} \right|^r dx \right)^{\frac{1}{r}} \left( \int_{S_P} \varrho^{(1-n)\frac{r}{r-1}} \, dx \right)^{\frac{r-1}{r}}$$

by Hölder's inequality,

$$\leq \text{constant} \cdot s^{n+1-\frac{n}{r}} \left( \int_{S_P} \left| \frac{\partial u}{\partial \varrho} \right|^r d\, x \right)^{\frac{1}{r}}.$$

A similar estimate holds for the term $\int_S | u(R) - u(Q) | \, d\, R$, and the result follows.

We return now to functions defined in the full n-space.

Suppose $r < n$. We shall prove a stronger formulation of (2.2), namely

$$(2.4) \qquad | u |_{\frac{nr}{n-r}} \leq \frac{r}{2} \frac{n-1}{n-r} \prod_i \left| \frac{\partial u}{\partial x_i} \right|_r^{\frac{1}{n}}.$$

For $1 < r < n$ (2.4) follows from the special case $r = 1$, as one readily verifies, by simply applying the inequality for $r = 1$ to the function $V = | u |^{\frac{n-1}{n-r} r}$ and using Hölder's inequality in a suitable way. Thus it suffices to prove (2.4) for the case $r = 1$.

$$(2.4)' \qquad | u |_{\frac{n}{n-1}} \leq \frac{1}{2} \prod_i \left| \frac{\partial u}{\partial x_i} \right|_1^{\frac{1}{n}}.$$

We shall prove (2.4)′ here for $n = 3$. One sees easily that

$$| u(x) | \leq \frac{1}{2} \int_i \left| \frac{\partial u}{\partial x_i} \right| d\, x_i \qquad i = 1, 2, 3.$$

where $\int_i$ denotes integration along the full line through $x$ parallel to the $x^i$, axis. Thus

$$| 2\, u(x) |^{\frac{3}{2}} \leq \left( \int_1 \left| \frac{\partial u}{\partial x_1} \right| d\, x_1 \right)^{\frac{1}{2}} \cdot \left( \int_2 \left| \frac{\partial u}{\partial x_2} \right| d\, x_2 \right)^{\frac{1}{2}} \cdot \left( \int_3 \left| \frac{\partial u}{\partial x_3} \right| d\, x_3 \right)^{\frac{1}{2}}.$$

Integrating with respect to $x_1$ then $x_2$, and then $x_3$ we find with the aid of Schwarz's inequality

$$\int_1 | 2\, u(x) |^{\frac{3}{2}} \, d\, x_1 \leq \left( \int_1 \left| \frac{\partial u}{\partial x_1} \right| d\, x_1 \right)^{\frac{1}{2}} \cdot \left( \iint_{12} \left| \frac{\partial u}{\partial x_2} \right| d\, x_2 \, d\, x_1 \right)^{\frac{1}{2}} \cdot \left( \iint_{13} \left| \frac{\partial u}{\partial x_3} \right| d\, x_3 \, d\, x_1 \right)^{\frac{1}{2}}.$$

$$\iint_{21} | 2u |^{\frac{3}{2}} \, dx_1 dx_2 \leq \left( \iint_{21} \left| \frac{\partial u}{\partial x_1} \right| dx_1 \, dx_2 \right)^{\frac{1}{2}} \cdot \left( \iint_{12} \left| \frac{\partial u}{\partial x_2} \right| dx_2 \, dx_1 \right)^{\frac{1}{2}} \cdot \left( \iiint \left| \frac{\partial u}{\partial x_3} \right| dx \right)^{\frac{1}{2}},$$

and finally

$$\iiint |2u|^{\frac{3}{2}} d x \leq \left(\iiint \left|\frac{\partial u}{\partial x_1}\right| d x\right)^{\frac{1}{2}} \left(\iiint \left|\frac{\partial u}{\partial x_2}\right| d x\right)^{\frac{1}{2}} \left(\iiint \left|\frac{\partial u}{\partial x_3}\right| d x\right)^{\frac{1}{2}}.$$

that is, (2.4)′.

For general $n$ the inequality is proved in the same way with the aid of Hölder's inequality.

Suppose finally, for $j = 0$, $m = 1$, that $r = n$; this is the exceptional case 2. We claim that

$$|u|_p \leq \text{costant } |Du|_n^{\frac{1-q}{p}} |u|_p^{\frac{q}{p}} \quad 0 < q \leq p < \infty$$

where the constant depends only $n$, $q$ and $p$. It suffices to show this for large $p$ and this is easily done by appliyng (2.4)′ to the function $v = |u|^{p(1-1/n)}$, and using Hölder's inequality in a judicions manner.

Let us now consider the other extreme case $a = j/m$. It suffices to consider the case $j = 1$, $m = 2$, the general case may then be proved by induction on $m$. We claim that the following holds

$$(2.5) \qquad |Du|_p \leq c\, |D^2 u|_r^{\frac{1}{2}} |u|_q^{\frac{1}{2}} \quad \text{for} \quad \frac{2}{p} = \frac{1}{r} + \frac{1}{q}; \; 1 \leq q, r \leq \infty,$$

*with c an absolute constant.* Incidentally, as Ungar pointed out, we may permit $q$ to be any positive number, but $I$ shall confine myself to the case cited, in fact to the case $q$ finite, $1 < r < \infty$. The general case may be obtained by a slightly different argument, or just by letting $q$ tend to $\infty$, and $r$ tend to 1 or $\infty$ in (2.5).

Inequality (2.5) follows from the corrisponding inequality in one dimension

$$(2.6) \qquad \int |u_x|^p d x \leq c^p \left(\int |u_{xx}|^r d x\right)^{\frac{p}{2r}} \left(\int |u|^q d x\right)^{\frac{p}{2q}}, \quad \frac{2}{p} = \frac{1}{r} + \frac{1}{q},$$

which holds for the full, or half-infinite line (with $c$ an absolute constant), by integrating with respect to the other variables and appliyng Hölder's inequality.

Our proof of (2.6), though elementary, is slightly tricky. Peter Ungar has found another slightly longer proof which furnishes a better value for $c$.

The proof is based on a simple lemma which we leave as an exercise.

**Lemma**: *On an interval $\lambda$, whose lenght we also denote by $\lambda$, we have*

$$(2.7)\qquad \int_\lambda |u_x|^p\,dx \leq \bar{c}^p \lambda^{1+p-\frac{p}{r}} \left(\int_\lambda |u_{xx}|^r\,dx\right)^{\frac{p}{r}} + \bar{c}^p \lambda^{-\left(1+p-\frac{p}{r}\right)} \left(\int_\lambda u^q\,dx\right)^{\frac{p}{q}}$$

*with $\bar{c}$ an absolute constant.*

We shall prove that for any interval $L: 0 \leq x \leq L$ the following inequality holds

$$(2.8)\qquad \int_0^L |u_x|^p\,dx \leq 2\bar{c}^p \left(\int_0^\infty |u_{xx}|^r\,dx\right)^{\frac{p}{2r}} \left(\int_0^\infty |u|^q\,dx\right)^{\frac{p}{2q}}.$$

(2.6) follows easily from (2.8).

In proving (2.8) we may suppose that $|u_{xx}|_r = 1$. We shall cover the interval $L$ by a finite number of successive intervals $\lambda_1, \lambda_2, \ldots$, each one having as initial point the end point of the preceding. For $k$ a fixed positive integer, choose first the interval $\lambda: 0 \leq x \leq \frac{L}{k}$, and consider (2.7) for this interval. If the first term on the right of (2.7) is greater than the second set $\lambda_1 = \lambda$; we then have

$$\int_{\lambda_1} |u_x|^p\,dx \leq 2\bar{c}^p \left(\frac{L}{k}\right)^{1+p-\frac{p}{r}},$$

since $|u_{xx}|_r = 1$. If however the second term of (2.7) is the greater extend the interval $\lambda$ (keeping its left endpoint fixed) until the two terms of the right of (2.7) become equal. Since $1 + p - \frac{p}{r} > 0$ equality of these two terms must occur for a finite value of $\lambda$. Let $\lambda_1$ be the resulting interval. We then have

$$\int_{\lambda_1} |u_x|^p\,dx \leq 2\bar{c}^p \left(\int_{\lambda_1} |u_{xx}|^r\,dx\right)^{\frac{p}{2r}} \left(\int_{\lambda_1} |u|^q\,dx\right)^{\frac{p}{2q}}.$$

Starting at the end point of $\lambda_1$ repeat this process, keeping $k$ fixed, choosing $\lambda_2, \lambda_3, \ldots$, until $L$ is covered. There are clearly at most $k$ such intervals $\lambda_j$. If we now sum our estimates for $\int_{\lambda i} |u_x|^p\,dx$ we find, with the aid of Hölder's inequality (recall that $\frac{p}{2r}+\frac{p}{2q}=1$) that

$$\int_0^L |u_x|^p\,dx \le 2\,\bar{c}^p\left(\frac{L}{k}\right)^{1+p-\frac{p}{r}}\cdot k + 2\,\bar{c}^p\left(\int_0^\infty |u_{xx}|^r\,dx\right)^{\frac{p}{2r}}\cdot$$

$$\cdot\left(\int_0^\infty |u|^q\,dx\right)^{\frac{p}{2q}}.$$

If we now let $k\to\infty$ the first term on the right of the preceding tends to zero, because $r>1$, and we obtain (2.8), completing the proof of (2.5).

## Lecture III. The Dirichlet Problem.

We consider now elliptic differential operators, confining ourselves for simplicity to a single equation for one unknown. Let $L(x, D)$ be a partial differential operator with complex valued coefficients, and let $L'$ be the part of highest order. $L$ is elliptic if there are no real characteristics, i. e.,

$$(3.1)\qquad L'(x,\xi)\neq 0, \qquad \text{real}\quad \xi\neq 0.$$

It is easily seen that for more than two variables, $n>2$, ellipticity implies that the order $k$ of $L$ is even. In treating the Dirichlet problem we shall assume that $k=2m$ is even and that the operator is strongly elliptic, i. e. that (efter possibly multiplying by a suitable complex function)

$$(3.2)\qquad \Re e\, L'(x,\xi)\neq 0, \qquad \text{real}\quad \xi\neq 0.$$

The Dirichlet problem consists of finding a solution in a domain $\mathcal{D}$ of

$$Lu=f \qquad \text{in}\quad \mathcal{D}$$

$$\left(\frac{\partial}{\partial \vec{n}}\right)^j u=\Phi_j \qquad \text{on}\quad \dot{\mathcal{D}}, \qquad j=0,\dots,m-1,$$

where $\partial/\partial\vec{n}$ represents differentiation normal to the boundary. Here $f$ and $\Phi_j$ are given functions in $\mathcal{D}$ and $\dot{\mathcal{D}}$ respectively.

We shall describe here the Hilbert space approach to the Dirichlet problem, which is based on some form of the projection theorem, and is related to the classical method of minimizing the Dirichlet integral. In its present form the existence theory is mainly due to Gårding, Vishik, Browder and others; we refer the reader to [9] and [C] for expositions and references. This and the following lecture comprise a brief description of [9]. The theory is based on a single $L_2$ inequality. Gårding's inequality, expressing the positive definiteness of the Dirichlet integral associated with the differential operator.

Since this approach to the Dirichlet problem requires considerable differentiability assumptions on the coefficients we shall assume for simplicity that they are of class $C^\infty$ in $\overline{\mathcal{D}}$ and that the boundary $\dot{\mathcal{D}}$ is sufficiently smooth. We shall also assume $\mathcal{D}$ to be bounded. Furthermore if the $\Phi_j$ are sufficiently smooth we may subtract from $u$ a function having the same Dirichlet data as $u$, so we shall consider the case where the $\Phi_j$ vanish

$$(3.3) \qquad Lu = f \quad \text{in } \mathcal{D}$$

$$\left(\frac{\partial}{\partial \vec{n}}\right)^j u = 0 \qquad \text{on } \dot{\mathcal{D}}, \qquad j = 0, \dots, m-1.$$

The Hilbert space approach yields at first « generalized solutions » of (3.3) which we must define. A function $u$ which belongs, say, to $L_2$ in every compact subdomain of $\mathcal{D}$ is a « weak » solution of $Lu = f$ if

$$(3.4) \qquad (u, L^* \varphi) = (f, \varphi)$$

for every $\varphi$ which belongs to $C_0^\infty(\mathcal{D})$, i.e. is of the class $C^\infty$ and has compact support in $\mathcal{D}$. Here ( , ) denotes the $L_2$ scalar product, and $L^*$ is the formal adjoint of $L$. In addition to the $L_2$ norm we also introduce the Hilbert spaces $H_j(\mathring{H}_j)$, $j$ a non-negative integer. These are the closures in the norm (using the notation of Lecture I)

$$\| u \|_j = \left[ \sum_{|\beta| \le j} \int_{\mathcal{D}} | D^\beta u |^2 \, dx \right]^{\frac{1}{2}}$$

of the spaces $C^\infty(\mathcal{D})\ C_0^\infty(\mathcal{D})$. The associated norm and spaces relative to a subdomain $\mathcal{a}$ will be denoted by $\| \ \|_j^{\mathcal{a}}, H_j^{\mathcal{a}}, \mathring{H}_j^{\mathcal{a}}$. Clearly $H_0 = \mathring{H} = L_2$.

We remark that for $j > i$ $H_j \subset H_i$ and the set $\| u \|_j \le$ constant is compact in $H_i$.

Following Sobolev and Friedrichs we say that a function $u$ in $\mathcal{D}$ has strong derivatives in $L_2$ up to order $j$ in $\mathcal{D}(\overline{\mathcal{D}})$ if $u$ belongs to $H_j^{\mathcal{a}}(H_j^{\mathcal{D}})$ for every compact subdomain $\mathcal{a}$ of $\mathcal{D}$. With the aid of the results of the preceding lecture we see that a function in $H_j$ is continuous if $2j > n$. Functions in $\mathring{H}_m$ satisfy the boundary conditions of (3.3) in a generalized sense.

We now formulate the

GENERALIZED DIRICHLET PROBLEM: *Given $f$ in $H_0$ find a weak solution $u$ in $\mathring{H}_m$ of $Lu = f$.*

Using the notation of Lecture 1 we may write the operator $L$ in the form

$$L = \sum_{|\beta|, |\gamma| \leq m} D^\beta a_{\beta,\gamma} D^\gamma .$$

If $u$ is a weak solution in $\mathring{H}_m$ we may then carry out some partial integration in equation (3.4) and write it as

$$(3.4)' \qquad B[u, \varphi] = \Sigma (-1)^{|\beta|} (a_{\beta,\gamma} D^\gamma u, D^\beta \varphi) = (f, \varphi), \ \varphi \ \text{ in } C_0^\infty(\mathcal{D}) .$$

$B[u, v]$ is linear in $u$, antilinear in $v$ and satisfies, by Schwarz'inequality

$$(3.5) \qquad | B[u, v] | \leq \text{constant} \| u \|_m \cdot \| v \|_m .$$

We shall assume the strong ellipticity (3.2) to hold uniformly, i. e. for some positive constant $c_0$

$$\mathcal{Re}\, (-1)^m \sum_{|\beta|, |\gamma| = m} a_{\beta,\gamma}(x)\, \xi^\beta \xi^\gamma \geq c_0 | \xi |^{2m}, \qquad \xi \text{ real,}$$

for all $x$ in $\overline{\mathcal{D}}$. Our main result is

THEOREM: *For $\overline{C}$ sufficiently large the generalized Dirichlet problem for the equation $(L + \overline{C}) u = f$ admits a unique solution. For the equation $Lu = f$ we have the Fredholm alternative.*

The $L_2$ estimate on which the theorem is based is

GÅRDING'S INEQUALITY: *There exist constants $c > 0$ and $C$ such that*

$$(3.6) \qquad \mathcal{Re}\, B[\varphi, \varphi] = \mathcal{Re}\, (L\varphi, \varphi) \geq c \| \varphi \|_m^2 - C \| \varphi \|_0^2$$

*holds for every $\varphi$ in $C_0^\infty(\mathcal{D})$.*

This will be proved in the next lecture. It is clear from (3.5) that the inequality extends also to functions in $\mathring{H}_m$, and it follows from (3.6) that the only solution in $\mathring{H}_m$ of $(L + C) u = 0$ is $u = 0$.

Let us now prove the theorem. Suppose first that the operator is symmetric, i. e. $B[\varphi, \varphi]$ is real. and that the constant $C$ in (3.6) vanishes — which we may achieve by considering $L + C$ in place of $L$. It follows from (3.5), (3.6) (with $C = 0$) that $B[u, v]$ serves as an alternative scalar product in the Hilbert space $\overset{\circ}{H}_m$; the norms $B[u, u]$ and $\| u \|_m$ are equivalent. We see that the antilinear functional $(f, \varphi)$ defined for all $\varphi$ in $\overset{\circ}{H}_m$ satisfies

$$|(f, \varphi)| \leq \|f\|_0 \|\varphi\|_0 \leq \|f\|_0 \|\varphi\|_m \leq c^{-\frac{1}{2}} \|f\|_0 B[\varphi, \varphi]^{\frac{1}{2}}$$

and is therefore a bounded functional. By the well known representation theorem there exists therefore a function $u$ in the Hilbert space $\overset{\circ}{H}_m$ such that

$$(f, \varphi) = B[u, \varphi];$$

$u$ is then the solution of the Dirichlet problem, and we have proved the first part of the theorem with $\bar{C} = C$. To prove the second part we write the equation $Lu = f$ in the form $(L + C) = Cu + f$ or

$$u = C(L + C)^{-1} u + (L + C)^{-1} f.$$

Since $(L + C)^{-1}$ maps $H_0$ boundedly into $\overset{\circ}{H}_m$ it is completely continuous in $H_0$, by a previous remark, and from the Riesz theory for completely continuous operators we derive the second part of the theorem.

Suppose now that $B[\varphi, \varphi]$ is not symmetric. If we add $C(\varphi, \varphi)$ to $B$ so that it satisfies

$$B[\varphi, \varphi] \geq c \|\varphi\|_m^2, \qquad \varphi \in \overset{\circ}{H}_m,$$

then we may still rely on a generalized representation theorem due to Lax and Milgram. We conclude the lecture with this

REPRESENTATION THEOREM: *Let $B(x, y)$ be a form defined for pairs of vector $x, y$ in a Hilbert space $H$* (norm $\| \;\|$), *which is linear in $x$, antilinear in $y$, and satisfies*

$$(3.7) \qquad |B(x, y)| \leq \text{constant} \|x\| \cdot \|y\|.$$

*Suppose that for some positive constant $c$ the inequality*

$$(3.8) \qquad |B(x, x)| \geq c \|x\|^2$$

*holds for every $x$ in $H$. Then every bounded antilinear functional $F(x)$ admits the representation*

$$F(x) = B(v, x) = \overline{B(x, w)}.$$

*For fixed elements $v, w$ which are unique.*

*Proof*: For any fixed element $v$, $B(v, x)$ is a bounded antilinear functional of $x$ and therefore admits the representation

$$B(v, x) = (y, x)_H$$

for some element $y$, where $(\ ,\ )_H$ denotes the scalar product in $H$. This defines a mapping $y = Av$ which is clearly linear. Letting $x = v$ and applying (3.8) we find that

$$c\,|v|^2 \leq |B(v, v)| \leq (y, v)_H \leq \|y\| \cdot \|v\|,$$

or

$$\|v\| \leq c^{-1}\,\|y\|.$$

It follows that the operator $A$ has a bounded inverse and that its range is closed. Furthermore the $v$ corresponding to any $y$ is unique. To see that the range of $A$ is the whole space $H$ suppose that $z$ is orthogonal to it. Then we have $B(v, z) = 0$ for all $v$. From (3.8) it follows, by setting $v = z$, that $z = 0$. Thus $A$ maps onto the entire space, and therefore every antilinear functional $F(x)$ being of the form $(y, x)_H$ admits the representation $F(x) = B(v, x)$. The other representation is proved in a similar way.

## Lecture IV. A Priori Estimates.

Before proving Garding's inequality let us make some general remarks about a priori estimates. Consider a differential equation $Lu = f$ of order $k$ and assume that the solution has been made unique by some auxiliary conditions. One wants to study the inverse operator — to see, for instance, to what class of functions the solution belongs, if $f$ belonge to a given class. For this problem, and also for the existence theory, a priori inequalities play a basic role. Let us suppose that the auxiliary conditions are homogeneous, then a typical a priori estimate would assert that for some norm $\|\ \ \|$

$$\|D^\beta u\| \leq \text{constant}\,\|Lu\| \qquad |\beta| \leq K.$$

For instance, if we know that the equation has a solution of class $C^K$ for all $f$ of class $C^j$ then indeed, by a simple application of the closed graph theorem, we would have

$$\|D^\beta u\| \leq \text{constant}\,\|Lu\|, \qquad |\beta| \leq K - j,$$

with $\| \quad \|$ the usual norm in $C^j$. In general if $Lu$ has finite $\| \quad \|$ norm we will not obtain such an inequality for $K = k$, rather $K < k$; that is we cannot estimate individually all derivatives entering in $L$. However I believe that elliptic equations can be characterized as those for which one can estimate all derivatives, i. e.

$$(4.1) \qquad \| D^\beta u \| \leq \text{constant} \, \| Lu \| , \qquad | \beta | \leq k ,$$

for a wide class of norms (this is stated as a conviction not a theorem).

Consider now an elliptic equation $Lu = f$ with suitable homogeneous boundary conditions. Most a priori estimates are just of the type (4.1) or, if one does not assume uniqueness, of the form

$$(4.2) \qquad \| D^\beta u \| \leq \text{constant} \, ( \| Lu \| + \| u \| ) \qquad | \beta | \leq k .$$

Indeed much of the theory of elliptic equations is concerned with proving such estimates for various norms $\| \quad \|$, and proving analogous estimates for functions with no boundary restrictions:

$$(4.2)' \qquad \| D^\beta u \|^{\mathcal{C}} \leq \text{constant} \, ( \| L \|^{\mathcal{D}} + \| u \|^{\mathcal{D}} ) \qquad | \beta | \leq k .$$

Here $\mathcal{C}$ is any compact subdomain of $\mathcal{D}$, and the norm $\| \quad \|^{\mathcal{C}}$ is considered only for functions defined in $\mathcal{C}$.

*A word of caution*: The estimate do not hold for the most obious norm that one would try, namely the maximum (or $C^0$) norm nor in general for $C^j$ norms, however they do hold for $C^{j+\alpha}$ norms, $0 < \alpha < 1$, and for many integral norms.

We quote some immediate consequence of (4.2), (4.2)′.

1. If $f$ and the coefficients of $L$ are in $C^\infty$ then a solution of $Lu = f$ is also in $C^\infty$. This follows fairly easily from (4.2)′.

2. Solutions of $Lu = 0$ with bounded norm $\| \quad \|$ form a compact family. This follows from (4.2)′ and the

*Calculus Lemma: The set* $\| u \| + \| Du \|$ constant *is compact in the space with* $\| \quad \|$ *as norm.*

This lemma holds for a wide class of norms.

3. The set of solutions of $Lu = 0$ satisfying the boundary conditions (so that (4.2) holds) is finite dimensional. This follows with the aid of the Calculus Lemma.

I would like to describe briefly a general recipe for proving such estimates. This consists of several steps:

1. In case of (4.2)′ prove it first for equations with constant coefficients and only highest order terms, and for functions of compact support.

In case (4.2), prove it also for such equations and for functions defined in a half space, vanishing near infinity, and satisfiyng (on the planar boundary) the boundary conditions. These are also assumed to have constant coefficients (i. e. to be translation invariant).

2. Now eliminate the hypothesis of compact support.

3. Extend the estimate to variable coefficients as follows: with the aid of a partitions of unity write the function $u$ as a sum of functions $u_i$ with small support, in each of which the leading coefficients are close to constants, and treat the variation from constant as an error term, using the results of Step 2 and the following lemma which may also be used in the proof of Step 2.

*Calculus Lemma: For appropriate constants* $c_1$, $c_2$

$$\| D^j u \| \leq c_1 \| D^m u \|^{j/m} \| u \|^{1-j/m} + c_2 \| u \| , \tag{4.3}$$

*where for functions of compact support we may take* $c_2 = 0$ *and* $c_1$ *independent of the support of* $u$.

This holds for a wide class of norms.

In case the support of $u_i$ touches the boundary, make a local change of variable to flatten out the boundary so that Steps 1 and 2 can be applied.

The main step here is Step 1. We remark that in Step 3 we rely on (at least) the continuity of the leading coefficients of $L$, or on the fact that they differ little from constants in small domains. Because of this one does not obtain in this way the more refined estimates required for treating nonlinear problems, such as those in Bers, Nirenberg [10], de Giorgi [11], or Nash [12].

The norms for which such estimates are easiest to derive are the $L_2$ norms for functions and their derivatives, and we shall illustrate the recipe for these by proving Gårding's inequality in its general form.

Consider a quadratic integral form defined for $C^\infty$ functions with compact support in a bounded domain $\mathcal{D}$

$$B[u, u] = \sum_{|\beta|, |\gamma| \leq m} (c_{\beta,\gamma} D^\beta u, D^\gamma u) \tag{4.4}$$

and suppose that the (complex valued) coefficients $c_{\beta,\gamma}$ are continuous in $\overline{\mathcal{D}}$. *A necessary and sufficient condition for the existence of positive constants* $c$, $C$ *so that the inequality*

$$\| u \|_m^2 \leq c \,\mathcal{R}e\, B[u, u] + C \| u \|_0^2 \tag{4.5}$$

***holds for all*** $u \in C_0^\infty(\mathcal{D})$ ***is that for some positive constant*** $c_0$

$$\Re e \sum_{|\beta|, |\gamma| = m} c_{\beta,\gamma}\, \xi^\beta \xi^\gamma \geq c_0 \, | \xi |^{2m} \quad \text{for all real } \xi . \tag{4.6}$$

**Here the notations of Lecture 3 is used.**

***Proof*: We prove first the sufficiency, following our recipe. The Calculus Lemma (4.3) will be used in the form: For every** $\varepsilon > 0$ **there is a constant** $C(\varepsilon)$ **such that for every** $C^\infty$ **function** $u$ **with compact support**

$$\| u \|_{m-1}^2 \leq \varepsilon \, \| u \|_m^2 + C(\varepsilon) \, \| u \|_0^2 . \tag{4.7}$$

**This is contained in our inequalities of Lecture 2, but is most easily proved with the aid of Fourier transforms.**

**We consider now the different steps in proving (4.5), the Step 2 of the recipe does not occur here since our functions have compact support.**

**1. Suppose that the** $c_{\beta,\gamma}$ **are constant and vanish unless** $|\beta| = |\gamma| = m$. **We introduce the Fourier transform of** $u$

$$\tilde{u}(\xi) = \int e^{-ix\cdot\xi} \, u(x) \, dx .$$

**By Parseval's theorem we have**

$$\Re e \, B[u, u] = (2\pi)^{-n} \, \Re e \sum \int c_{\beta,\gamma} \, \xi^\beta \xi^\gamma \, | \tilde{u}(\xi) |^2 \, d\xi$$

$$\geq (2\pi)^{-n} c_0 \int | \xi |^{2m} \, | \tilde{u}^2 | \, d\xi$$

$$\geq c_0' \, \| u \|_m^2$$

**proving (4.5) for this special case.**

**We now consider the variable coefficient case and break Step 3 into two parts.**

**2. Suppose that the support of** $u$ **is sufficiently small, contained, say, in a small sphere about the origin. Then accorfing to the preceding inequality we have**

$$c_0' \, \| u \|_m^2 \leq \Re e \, B[u, u] + \Re e \sum_{|\beta| = |\gamma| = m} \int (c_{\beta,\gamma}(0) - c_{\beta,\gamma}(x)) \, D^\beta u \, D^\gamma \bar{u} \, dx \, -$$

$$- \, \Re e \sum_{|\beta| + |\gamma| < 2m} \int c_{\beta,\gamma}(x) \, D^\beta u \, D^\gamma \bar{u} \, dx .$$

If now the support of $u$ is so small that $c_{\beta,\gamma}$ has small oscillation there we see that the second term on the right may be bounded by

$$\frac{1}{2} c_0' \| u \|_m^2 .$$

The third term is trivially bounded by constant $\| u \|_m \| u \|_{m-1}$.

Thus we find that

$$\frac{1}{2} c_0' \| u \|_m^2 \leq \mathfrak{Re}\, B[u, u] + \text{constant} \| u \|_m \| u \|_{m-1}$$

from which follows the inequality

$$\| u \|_m^2 \leq \text{constant}\, \mathfrak{Re}\, B[u, u] + \text{constant} \| u \|_{m-1}^2 .$$

(4.5) now follows with the aid of (4.7).

3. Consider now the general case. Construct a partition of unity in $\overline{\mathcal{D}}$,

$$1 \equiv \sum_1^N \omega_j^2 , \qquad \omega_j \in C_0^\infty ,$$

with the support of each $\omega_j$ as small as desired. Then

$$\mathfrak{Re}\, B[u, u] = \mathfrak{Re} \sum \int c_{\beta,\gamma} D^\beta u\, D^\gamma \overline{u}\, dx =$$

$$= \mathfrak{Re} \sum \sum_j \int \omega_j^2\, c_{\beta,\gamma} D^\beta u\, D^\gamma \overline{u}\, dx =$$

$$= \mathfrak{Re} \sum \sum \int c_{\beta,\gamma} D^\beta (\omega_j u)\, D^\gamma (\omega_j \overline{u})\, dx + 0 (\| u \|_m \cdot \| u \|_{m-1})$$

$$\geq \text{constant} \sum_{|\beta| \leq m} \sum_j \int | D^\beta (\omega_j u) |^2\, dx + + 0 (\| u \|_m \cdot \| u \|_{m-1})$$

by the preceding Case 2,

$$\geq \text{constant} \| u \|_m^2 + 0 (\| u \|_m \cdot \| u \|_{m-1}) ,$$

and the desired result now follows easily with the aid of (4.7).

We see that the constants $c$, $C$ in (4.5) depend on $c_0$, an upper bound for the $|c_{\beta,\gamma}|$, and on the modulus of continuity of the leading $c_{\beta,\gamma}$ with $|\beta| = |\gamma| = m$, and finally on the domain $\mathcal{D}$.

Now for the proof of the nacessity of (4.6). Suppose that (4.5) holds and that the left hand side of (4.5) vanishes for some real $\xi = \xi'$, $|\xi|' = 1$, and some point in $\overline{\mathcal{D}}$, say the origin. Following the ergument in Step 2 in the proof of sufficiency we see that the inequality

$$(4.5)' \qquad \| u \|_m^2 \leq \text{constant} \left( \mathcal{R}e \sum_{|\beta|,|\gamma|=m} \int c_{\beta,\gamma}(0)\, D^\beta u\, D^\gamma \bar{u}\, dx + \| u \|_0^2 \right)$$

holds for all $C^\infty$ $u$ with support in some fixed neighborhood $U$ about the origin and in $\mathcal{D}$. Set $u = e^{i\lambda\xi'\cdot x}\zeta(x)$ for real $\lambda$, where $\zeta(x)$ is a fixed real $C^\infty$ function with support in $U$ and in $\mathcal{D}$. One sees readily that as $\lambda \to \infty$ the left hand side of (4.5)′ is $0(\lambda^{2m})$ and not $o(\lambda^{2m})$ while the right hand side is $0(\lambda^{2m-1})$, so that (4.5)′ does not hold.

Gårding's inequality (4.5) is at one end of a whole spectrim of interesting and useful inequalities making different requirements on $u$ at the boundary, Gårding's inequality making the maximal requirement — that all derivatives of $u$ of order less than $m$ vanish at the boundary. At the other end of the spectrum is the inequality of Aronszajn [13] involving no boundary conditions whatsoever.

Aronszajn considers a number of differential operators $L_j(x, D)$, $j = 1, \ldots, N$ of order $m$, with coefficients continuous in the closure of a bounded domain $\mathcal{D}$, and solves the following problem: Under what conditions can one assert that for all $C^\infty$ functions $u$ in $\mathcal{D}$ the inequality

$$(4.8) \qquad \| u \|_m^2 \leq \text{constant}\, (\Sigma \| L_j u \|_0^2 + \| u \|_0^2)$$

holds, with the constant independent of $u$? He gives necessary and sufficient conditions:

(a) the operator $\Sigma L_j L_j^*$ is elliptic, here $L_j^*$ is the formal adjoint of $L_j$.

(b) At any boundary point $x$ of $\mathcal{D}$, if $\vec{n}$ is the unit normal to $\dot{\mathcal{D}}$ and $\xi \neq 0$ is any real vector tangent to $\dot{\mathcal{D}}$ then the polynomials in $\tau$, $L_j'(x, \xi + \tau\vec{n})$ have no common complex root $\tau$. Here $L_j'$ is the leading part of $L_j$.

An example of Aronzsajn's inequality is the following; for functions $u(x, y)$ in a bounded domain in the plane

$$\int |u_{xy}|^2\, dx\, dy \leq \text{constant} \int (|u_{xx}|^2 + |u_{yy}|^2 + |u|^2)\, dx\, dy.$$

Even this simple example is not trivial to prove.

Since the report of Aronszajn a number of people have coonsidered the problem of proving (4.5) for various quadratic forms (4.4) and under various differential boundary conditions. For one operator $L_j$ Agmon, Douglis, Nirenberg [14], (in a fortheoming paper which will be discussed later) have characterized these differential boundary conditions which are $m/2$ in number and for which (4.8) holds. Schechter [15] has treated $N$ operators and general boundary conditions. Aronszajn, in unpublished work, has treated the general problem (4.5). Also Hörmander and Agmon [16] have solved the general problem for (4.5) and general differential boundary conditions. The proofs follow the recipe outlined above, the main step being the first, for functions in a half space.

We conclude the lecture with a result that will be used in proving the differentiability at the boundary of solutions of elliptic equations. In the following $\Sigma_R$ denotes the hemisphere $|x| < R$, $x_n \geq 0$. We shall denote the variable $x_n$ by $t$, $(x_1, \dots, x_{n-1})$ by $x$ and $(x_1, \dots, x_n)$ by $(x, t)$.

***Lemma***: *Let $u$ be a weak solution of a differential equation (of order $k$) with, for simplicity, $C^\infty$ coefficients,*

$$Lu = \sum_{|\beta| \leq k-j-1} D^\beta f_j \tag{4.9}$$

*in the interior of a hemisphere $\Sigma_R$, where $f_\beta$ are given functions, and assume that the plane $t = 0$ is nowhere characteristic, in fact that the coefficient $a$ of $D_t^k$ in $L$ does not vanish. If for every $\delta > 0$ the functions $f_\beta$, $D^\beta u$ for $|\beta| \leq j$, $D_x D^j u$ belong to $L_2$ in $\Sigma_{R-\delta}$ then also the function $D_t^{j+1} u$ has this property.*

For $j \geq k-1$ there is nothing to prove, as we may solve for the function $D_t^{j+1} u$ from the differential equation (4.9) operated on by $D_t^{j+1-k}$. Thus we suppose $j < k-1$.

The proof makes use of a well known formula giving explicitly a smooth extension of a function $v$ defined in a half space $t > 0$ to a function defined in the full space:

$$v_N(x, t) = v(x, t) \qquad t > 0 \tag{4.10}$$

$$v_N(x, t) = \sum_{j=1}^{N} \lambda_j \, v(x, -j\,t) \qquad t < 0$$

with the $\lambda_j$ chosen so that

$$\sum_j (-j)^k \lambda_j = 1, \qquad k = 0, \dots, N-1.$$

We observe that,

$$\| v_N \|_k \leq \text{constant} \| v \|_k \qquad k = 0, \dots, N-1 .$$

Here the norm on the left is over the full space while on the right it is over the half space $t > 0$.

*Proof of the Lemma*: Choose a fixed $\delta > 0$, let $\zeta(x, t)$ be a fixed $C^\infty$ function with support in $|x|^2 + t^2 < R^2$ and which equals one in $\Sigma_{R-\delta}$, and set $\zeta a u = v$. If we can prove that $D_t^{j+1} v$ belongs to $L_2$ then, since $a \neq 0$ it follows easily that $D_t^{j+1} u$ is in $L_2$ in $\Sigma_{R-\delta}$. From our assumptions we see that $v$ is a weak solution of a differential equation of the form

$$D_t^k v = \sum_{s+|\gamma| \leq k-j-1} D_t^s D_x^\gamma v_{s,\gamma} \tag{4.11}$$

where the $v_{s,\gamma}$ belong so $L_2$, and that derivatives $D_x D^j v$ and $v$ itself belong to $L_2$.

For $N$ sufficiently large we now extend the functions $v, v_{s,\gamma}$ to negative $t$, defining $v_N$ by (4.10) and $v_{s,\gamma,N}$ by

$$v_{s,\gamma,N}(x, t) = v_{s,\gamma}(x, t), \qquad t > 0$$

$$v_{s,\gamma,N}(x, t) = \sum_{j=1}^{N} \lambda_j (-j)^{k-s} v_{s,\gamma}(x, -jt), \qquad t < 0 .$$

One may then verify that the equation

$$D_t^k v_N = \Sigma D_t^s D_x^\gamma v_{s,\gamma,N}$$

holds in the entire space in the weak sense, and that the $v_{s,\gamma,N}$, the derivatives $D_x D^j v_N$ and $v_N$ itself belong to $L_2$.

Let us now take Fourier transforms with respect to $x$ and $t$, and write $(\xi_1, \dots, \xi_{n-1}) = \xi$, $\xi_n = \tau$. Denoting the trasform of a function $f$ by $\tilde{f}$ we find that

$$(i\tau)^k \tilde{v}_N = \Sigma (i\tau)^s (i\xi)^\gamma \tilde{v}_{s,\gamma,N}, \tag{$\widetilde{4.11}$}$$

with $\tilde{v}_{s,\gamma,N}$, $\tilde{v}_N$ and $|\xi| (|\xi|^j + |\tau|^j) \tilde{v}_N$ belonging to $L_2$ in the $(\xi, \tau)$ space.

To conclude the proof we have to show that $\tau^{j+1} \tilde{v}_N$ belongs to $L_2$. To this end write

$$|\tau|^{j+1} \tilde{v}_N = \frac{|\tau|^{j+1}}{\tau^{2k} + |\xi|^{2k} + 1} \tau^{2k} \tilde{v}_N + \frac{|\tau|^{j+1}}{\tau^{2k} + |\xi|^{2k} + 1} (|\xi|^{2k} + 1) \tilde{v}_N . \tag{4.12}$$

We shall show that each term on the right belongs to $L_2$. From $(\widetilde{4.11})$ we find that the first term on the right is bounded by

$$\frac{|\tau|^{k+j+1}}{\tau^{2k}+|\xi|^{2k}+1}\Sigma|\tau|^s|\xi|^\gamma|v_{s,\gamma,N}|.$$

Since $s+|\gamma|\leq k-j-1$ it follows that the factor of $v_{s,\gamma,N}$ is uniformly bounded, and hence that this term belongs to $L_2$, since the $v_{s,\gamma,N}$ do.

The second term on the right of (4.12) is bounded by

$$c(|\xi|(|\xi|^j+|\tau|^j)+1)|\widetilde{v}_N|$$

with $c$ an absolute constant, and hence belongs also to $L_2$, by an earlier remark.

This completes the proof of the Lemma.

## Lecture V. The Differentiability of Weak Solutions of Elliptic Equations

In this and the next lecture we shall present a self contained proof of the well known result that solutions of elliptic equations with $C^\infty$ coefficients are of class $C^\infty$.

Many proofs exist in the literature including proofs for more general classes of equations, see Hörmander [17], Malgrange [18]. The proof here seems rather straigtforward; it is based essentially on a proof given by Lax [19] and is closely related to proofs given in lectures by Bers [20] and Schwartz [21] (see also [9]). We confine ourselves as before to a single equation (not necessarily strongly elliptic) although the argument extends also to systems.

*Differentiability Theorem: If $u$ is a locally square integrable weak solution of the elliptic equation $Lu=f$, and $f\in C^\infty$ then $u\in C^\infty$.*

*Remark:* If $u$ is a distribution solution then $u=\Delta^k v$ for some continuous $v$ (here $\Delta$ is the Laplace operator), and $v$ is then a weak solution of $L\Delta^k v=f$. The Theorem holds therefore for this case also.

The proof consists in showing that $u$ has $L_2$ derivatives of all orders in every compact subdomain. That $u\in C^\infty$ then follows from the Sobolev estimates proved in Lecture 2. However since we only need a very simple case of the Sobolev lemmas we present a separate proof of it here.

*Lemma (Sobolev): In a « smooth » domain $\mathcal{D}$ if $u$ has $L_2$ derivatives up to order $s$ in $\overline{\mathcal{D}}$ for $s > n/2$, then $u$ is continuous in $\mathcal{D}$.*

*In fact*

$$\max |u| \leq K \left( \int \sum_{j=0}^{s} | D^j u |^2 \, d x \right)^{1/2}, \qquad s > n/2 .$$

*Proof:* The first assertion follows easily from the inequality. To prove the inequality let $x_0$ be an inner point in $\mathcal{D}$ (for simplicity take $x_0 = 0$) and suppose there is a sphere about $x_0$ in $\mathcal{D}$ with radius $R$. Let furthermorl $\zeta(r)$ be a function in $C^\infty$, equal to 1 for $0 \leq r \leq R/2$, and vanishing for $r \geq R$. By integration along any radius from $x_0 = 0$, and by repeated partial integration we see that

$$u(0) = - \int_0^R 1 \cdot (\zeta u)_r \, d r = \text{const} \int_0^R r^{s-1} \left(\frac{\partial}{\partial r}\right)^s (\zeta u) \, p r .$$

integrating over the unit sphere (with area $\Omega$) of radial directions one finds

$$| \Omega u(0)) | = | \text{const} \int r^{s-n} \left(\frac{\partial}{\partial r}\right)^s (\zeta u) \, d x | \leq$$

$$\leq \text{const} \left( \int \left| \left(\frac{\partial}{\partial r}\right)^s (\zeta u) \right|^2 d x \right)^{1/2} \left( \int r^{2(s-n)} \, d x \right)^{1/2}$$

using Schwarz inequality. For $s > n/2$ the last integral is finite.

If the boundary of $\mathcal{D}$ is such that at any point in $\overline{\mathcal{D}}$ there exists a cone with a fixed opening and length contained in $\overline{\mathcal{D}}$ then the same proof holds; instead of integrating over the full sphere of radial directions, we merely integrate over the directions lying in the cone.

The proof of the Differentiability Theorem consists mainly of a series of simple lemmas of calculus concerned with a special situation) that of periodic functions, and this lecture will confined to these calculus statements.

We consider *periodic* functions $u \in C^\infty$ with period $2\pi$ in each $x_j$. For such functions the Fourier series

$$u = \sum_{\xi} u_\xi \, e^{i x \cdot \xi}, \quad \xi = (\xi_1 \ldots . \xi_n)$$

($\xi_j$ = integer) converges uniformly.

By Parseval's equality we have the following estimate for each non-negative $s$

$$(5.1) \quad \text{canstant} \sum_{\xi} (1 + |\xi|^2)^s |u_\xi|^2 \leqq \int \sum_{j=0}^{s} |D^j u|^2 \, dx$$

$$\leqq \text{constant} \sum_{\xi} (1 + |\xi|^2)^s |u_\xi|^2$$

where the integral is taken over a period cube.

For *any* integer $s$ we introduce the following scalar product and norm, differing slightly from our previous notation,

$$(u, v)_s = (2\pi)^n \sum_{\xi} (1 + |\xi|^2)^s u_\xi \overline{v_\xi}$$

$$\|u\|_s^2 = (u, u)_s .$$

We write $(u, u)_0 = (u, u)$ and proceed with the

*Calculus:*

1. $\|u\|_s$ is increasing in $s$. Furthermore for $t_1 < s < t_2$ and any $\varepsilon > 0$ there is a constant $C(\varepsilon)$ such that

$$(5.2) \qquad \|u\|_s \leq \varepsilon \|u\|_{t_2} + C(\varepsilon) \|u\|_{t_1} .$$

*Proof:* For any $\sigma \geq 0$, $\sigma^s \leq \varepsilon \sigma^{t_2} + C(\varepsilon) \sigma^{t_1}$.

2. Set $\varphi = (1 - \Delta)^t u$, $\psi = (1 - \Delta)^t v$, so that $\varphi = \sum_{\xi} u_\xi (1 + |\xi|^2)^t e^{ix\cdot\xi}$. From this we find

$$(5.3) \qquad \|u\|_s = \|\varphi\|_{s-2t} = \|(1 - \Delta)^t u\|_{s-2t}$$

$$(5.4) \qquad (u, v)_s = (u, (1 - \Delta)^t v)_{s-t} = ((1 - \Delta)^t u, v)_{s-t} .$$

As a consequence we have

*Lemma:* If $\omega \in C^\infty$, then

$$(5.5) \qquad (\omega u, v)_t = (u, \overline{\omega} v)_t + 0(\|u\|_t \|v\|_{t-1} + \|u\|_{t-1} \|v\|_t) .$$

*Proof:* Assume $t \leq 0$. Using (5.4), (5.3), (5.1), and partial integration, we find

$$(\omega u, v)_t = (\omega (1 - \Delta)^{-t} \varphi, \psi) = ((1 - \Delta)^{-t} \varphi, \overline{\omega} \psi)$$

$$= (\varphi, (1 - \Delta)^{-t} (\overline{\omega} \psi)) =$$

$$= (\varphi, \overline{\omega} (1 - \Delta)^{-t} \psi) + 0(\|\varphi\|_{-t} \|\psi\|_{-t-1} + \|\varphi\|_{-t-1} \|\psi\|_{-t})$$

$$= (u, \overline{\omega} v)_t + 0(\|u\|_t \|v\|_{t-1} + \|u\|_{t-1} \|v\|_t) .$$

In the case $t \geq 0$ the proof is similar.

3. *Schwar(t)z's inequality :*

(5.6) $$|(u, v)_s| \leq \| u \|_{s+t} \| v \|_{s-t} .$$ (Clear !)

In fact

(5.7) $$\| u \|_{s+t} = 1.u.b. \frac{|(u, v)_s|}{\| v \|_{s-t}} .$$

*Proof:* According to (5.6) the left side of (5.7) is not smaller than the right side. If however we set $v = (1 - \Delta)^t u$, then, by (5.4)

$$\frac{(u, v)_s}{\|v\|_{s-t}} = \frac{(u, (1 - \Delta)^t u)_s}{\| (1 - \Delta)^t u \|_{s-t}} = \frac{(u, u)_{s+t}}{\| u \|_{s+t}} = \| u \|_{s+t} ,$$

proving (5.7).

We can now form Hilbert espace $H_s$ by completing $C^\infty$ functions in the norms $\| \;\; \|_s$. For $s \geq 0$ these agree with our previous definitions. Obviously $H_s \subset H_t$ for $s > t$. All the previous results hold for functions with the appropriate norms finite, for instance (5.7). We may regard $H_s$ as given by a formal Fourier series with finite $\| \;\; \|_s$ norm.

We remark that the scalar product

$$(u, v)$$

is defined, by extension, for any functions $u \in H_s$, $v \in H_{-s}$, and that any bounded linear functional $f(u)$ defined on $H_s$ may be represented in the form

$$f(u) = (u, v)$$

with $v \in H_{-s}$; this follows immediately from the Fourier series representation, so that we may regard $H_{-s}$ as dual to $H_s$.

Though we shall not use this, we remark that the closed unit ball $\| u \|_s \leq 1$ in $H_s$ is compact in $H_t$ for $s > t$.

We continue with the calculus.

4. Consider any differential operator $L$ of order $k$ with $C^\infty$ coefficients. *Claim :*

(5.8) $$\| L u \|_s \leq \text{const} \| u \|_{s+k} .$$

More precisely

(5.9) $$\| L u \|_s \leq c K \| u \|_{s+k} + c K' \| u \|_{s+k-1} ,$$

where $c = c(k, n)$, $K$ is a bound for the leading coefficients, and $K'$ is a bound for all coefficients and their derivatives up to order $|s|$.

*Proof:* Since obviously $\| D^j u_s \| \leq \text{const} \| u \|_{s+j}$ it suffices, in order to prove (5.9), to show that if $a \in C^\infty$ then

$$\| a u \|_s \leq c k' \| u \|_s + c k'' \| u \|_{s-1} \tag{5.10}$$

where $k'$ and $k''$ are bounds for $| a |$ and $| D^j a | (j \leq | s |)$ respectively.

*Proof of* (5.10): Consider first the case $s < 0$. Set $\varphi = (1 - \Delta)^s u$ $\psi' = (1 - \Delta)^s$ an then we have, by (5.4), and partial integration,

$$\| a u \|_s^2 = \| \psi \|_{-s}^2 = (a u , \psi) = (a (1 - \Delta)^{-s} \varphi , \psi) .$$

Integrating the last by parts $(-s)$ times we find it is not greater than

$$c k \| \varphi \|_{-s} \| \dot{\psi} \|_{-s} + c k' \| \psi \|_{-s} \| \varphi \|_{-s-1} .$$

So dividing by $\| \psi \|_{-s}$ we have, with the aid of (5.3),

$$\| \psi \|_{-s} = \| a u \|_s \leq c k \| \varphi \|_{-s} + c k' \| \varphi \|_{-s-1}$$

$$= c k \| u \|_s + c k' \| u \|_{s-1} \text{ by (5.3).}$$

In case $s \geq 0$ we have

$$\| a u \|_s^2 = (a u , (1 - \Delta)^s a u) ,$$

and may integrate by parts as above.

So $L$ can be extended to all of $H_s$ and maps it boundedly into $H_{s-k}$. This operation of $L$ agrees with that of $L$ acting on $u$, regarded as a distribution.

*Technical Lemma:* Suppose $\omega$ is a $C^\infty$ real function, then

$$(L (\omega^2 u) , L u)_s = \| L(\omega u) \|_s^2 + 0 (\| u \|_{s+k} \| u \|_{s+k-1}) . \tag{5.11}$$

*Proof:*

$$(L (\omega^2 u) , L u)_s = (\omega L (\omega u) , L u)_s + 0 (\| u \|_{s+k} \| u \|_{s+k-1}) \text{ by (5.8)}$$

$$= (L (\omega u) , \omega L u)_s + 0 (\| u \|_{s+k} \| u \|_{s+k-1}) \text{ by (5.5)}$$

$$= (L (\omega u) , L (\omega u))_s + 0 (\| u \|_{s+k} \| u \|_{s+k-1}) \text{ by (5.8)}$$

To conclude this lecture we consider

*Difference Quotients:* For a given vector $h$ let

$$u^h = \frac{u (x + h) - u (x)}{| h |}$$

be the difference quotient. One verifies easily: $\| u(x+h) \|_s = \| u(x) \|_s$,

$$\| u^h \|_s \leq \| u \|_{s+1} . \tag{5.12}$$

Furthermore: If $u \in H_s$, $u^h \in H_s$ and $\| u^h \|_s \leq k$ for each $h$, then $\| u \|_{s+1} \leq k$.

*Corollary:* If $u \in H_s$, $\| u^h_s \| \leq k$ for each $h$, then $u \in H_{s+1}$ and $\| u \|_{s+1} \leq k$.

*Proof;* Let $u = \sum_{\xi} u_\xi e^{ix\cdot\xi}$, and let $u_N = \sum_{|\xi| \leq N} u_\xi e^{ix\cdot\xi}$. One finds $\| u_N \|_{s+1} \leq k$. Q. E. D.

## Lecture VI. Proof of the Differentiability Theorem.

Let now $L$ be an elliptic operator of order $k$. In the periodic case we prove the Differentiability Theorem in the form

DIFFERENTIABILITY THEOREM: *If* $u \in H_s$, $L u \in H_{s-k+1}$, *then* $u \in H_{s+1}$. *So it follows that if* $u \in H_s$ *and* $L u \in H_{t-k}$, *then* $u \in H_t$.

The non-periodic case is easily reduced to this as follows; We prove successively that $u$ has $L_2$ first order derivatives, then second order derivatives, then second order derivatives, and so on.

To carry out this reduction let $\zeta$ be a $C^\infty$ function defined in a neighborhood of a point and with compact support. Let $v = \zeta u$ and extend $v$ and the coefficients of $L$ as periodic functions. So

$$L v = L(\zeta u) = f + g$$

where $f = \zeta L u$, $g = L(\zeta u) - \zeta L u$; $g$ contains only derivatives of $u$ up to order $k-1$, and so, as is easily seen with aid of (5.8) has finite $\| \; \|_{1-k}$ norm.

So $L v \in H_{1-k}$, therefore $v \in H_1$ and so $u$ has $L_2$ derivatives in a neighborhood of the point. Using this one repeats the argument for a smaller neigborhood, and sees that $L v \in H_{2-k}$, so $v \in H_2$, and so on.

The proof of the Differentiability Theorem in the periodic case follows easily, in turn, from the following estimate which is the analogue of Gårding's inequality.

*Basic Estimate:* For any $s$, $s_0$

$$\| u \|_{s+k} \leq \text{constant} \; \| L u \|_s + \text{constant} \; \| u \|_{s_0} . \tag{6.1}$$

Postponing the proof of (6.1) let us prove the Differentiability Theorem. Consider a difference quotient $u^h$. If $L^h$ represents the operator obtained by replacing each coefficient in $L$ by its difference quotient we see that $Lu^h =$

$= (Lu)^h - L^h u(x+h)$ Thus we have, from (6.1)

$$\| u^h \|_s \leq \text{constant} \| L u^h \|_{s-k} + \text{constant} \| u^h \|_{s-1}$$

$$\leq \text{constant} \| (L u)^h \|_{s-k} + \text{constant} \| L^h u (x+h) \|_{s-k}$$

$$+ \text{constant} \| u \|_s \qquad \text{by (5.12)}$$

$$\leq \text{constant} \| L u \|_{s-k+1} + \text{constant} \| u \|_s \text{ by (5.12), (5.8)}$$

The desired result follows from the corollary after (5.12).

*Proof of the Basic Estimate:* Clearly only $s_0 < s + k$ is of interest. The proof consists of several steps, following our recipe, and the proof of Gårding's inequality in Lecture 4.

1. $L$ has constant coefficients with leading terms only. Then

$$\| L u \|_s^2 = (2\pi)^n \sum_\xi | u_\xi |^2 | L(\xi) |^2 (1 + |\xi|^2)^s$$

$$\geq \text{constant} \sum_\xi | u_\xi |^2 | \xi |^{2k} (1 + |\xi|^2)^s$$

while

$$\| u \|_{s_0}^2 = (2\pi)^n \sum_\xi | u_\xi |^2 (1 + |\xi|^2)^{s_0}.$$

Hence

$$\| L u \|_s^2 + \| u \|_{s_0}^2 \geq \text{constant} \sum_\xi | u_\xi |^2 (1 + |\xi|^2)^{s+k} =$$

$$= \text{constant} \| u \|_{s+k}^2.$$

2. $L$ has variable coefficients with leading coefficients differing from constant value by less than $\varepsilon$, $\varepsilon$ sufficiently small: Let $L_0$ be the operator with these constant coefficients. By case 1. we have

$$\| u \|_{s+k} \leq \text{constant} \| L_0 u \|_s + \text{constant} \| u |_{s_0}$$

$$\leq \text{constant} (\| L u \|_s + \| (L_0 - Lu) \|_s) + \text{constant} \| u \|_{s_0}$$

$$\leq \text{constant} \| L u \|_s + \text{constant}\, \varepsilon \| u \|_{s+k} + \text{constant} \| u \|_{s+k-1}, \text{ by (5.9)},$$

$$\leq \text{constant} \| L u \|_s + \text{constant}\, \varepsilon \| u \|_{s+k} + \frac{1}{2} \| u \|_{s+k} + \text{constant} \| u \|_{s_0} \qquad \text{by (5.2)},$$

from which (6.1) follows.

*Note:* If $u$ has its support in a small set then the leading coefficients, being continuous, difrer little from constant values. So (6.1) holds in that case.

3. General case: Intruduce a partition of unity over the closed period cube

$$1 = \sum_j \omega_j^2$$

with each $\omega_j$ having its support in a small region.

$$\| u \|_{s+k}^2 = (u, u)_{s+k} = (\sum_j \omega_j^2 u, u)_{s+k}$$

$$= \sum_j (\omega_j u, \omega_j u)_{s+k} + 0(\| u \|_{s+k} \| u \|_{s+k-1}) \text{ by } (5.5),$$

$$\leq \text{constant} \sum_j (L \omega_j u, L\omega_j u_s) + \text{constant} \sum_j \| \omega_j u \|_{s_0}^2 + 0(\| u \|_{s+k} \| u \|_{s+k-1})$$

(since $\omega_j u$ has its support in a small set)

$$= \text{constant } (L (\sum_j \omega_j^2 u)_0 L u)_s + 0(\| u \|_{s+k} \| u \|_{s+k-1}), \text{ by } (5.11),$$

$$\leq \text{constant} \| L u \|_s^2 + \frac{1}{2} \| u \|_{s+k}^2 + c \| u \|_{s+k-1}^2$$

$$\leq \text{constant} \| L u \|_s^2 + \frac{3}{4} \| u \|_{s+k}^2 + c \| u \|_{s_0}^2 -, \text{ by } (5.2)$$

from which (6.1) follows.

We conclude this lecture with some remarks concerning the differentiability near the boundary of the solution of the Dirichlet problem obtained in Lecture 3. Using the notation of that lecture we recall that the solution of $L u = f$ belonged to the space $H_m$. Since the discussion is local we may assume that the boundary is given by $x_n = 0$ and that we have a solution of the equation in the hemisphere $\Sigma_R$ (see Lecture 4). It suffices, by Sobolev, to show that the solution $u$ has derivatives of all orders in $L_2$ in $\Sigma_{R-\delta}$, for $f$ in $C^\infty$. We shall merely indicate the first step — the proof that $u$ has derivatives of order $(m+1)$ in $L_2$ (see [9]) — and shall use the notation of the lemma at the end of Lecture 4. By that lemma it suffices to prove that derivatives of the form $D_x D^m u$ belong to $L_2$ in $\Sigma_{R-\delta}$ for any $\delta > 0$.

This we do with the aid of difference quotients as above. For fixed $\delta > 0$ let $\zeta(x, t)$ be a $C^\infty$ function with support in $|x|^2 + t^2 < R^2$, and which equals one in $\Sigma_{R-\delta}$. Since the function $u$ satisfies

$$B[u, \varphi] = (f, \varphi)$$

for all $\varphi$ in $\mathring{H}_m(\Sigma_R)$ it follows that the function $v = \zeta u$ satisfies

$$B[v, \varphi] \leq \text{constant} \, \|\varphi\|_{m-1}$$

for such $\varphi$. If we nov form difference quotients $v^h$ as above with $h$ parallel to the boundary $t = 0$ we find easily that

$$B[v^h, \varphi] \leq \text{constant} \, \|\varphi\|_m$$

for $\varphi$ in $\mathring{H}_m(\Sigma_R)$. Setting $\varphi = v^h$ and applying Gårding's inequality for strongly elliptic operators of Lecture 3 we obtain a bound for

$$\|v^h\|_m$$

which is independent of $h$, and it follows easily that the derivatives $D_x D^m v$ are in $L_2$, hence that $D_x D^m u \in L_2$ in $\Sigma_{R-\delta}$.

## Lecture VII. A Priori Estimates Near the Boundary.

In the remainig time we shall discuss briefly the derivation of estimates near the boundary for solutions of elliptic equations in, for simplicity, a bounded domain $\mathcal{D}$. This material is taken from a paper by Agmon, Douglis, Nirenberg [14] which is concerned with both Schauder and $L_p$ estimates near the boundary for solutions satisfying general boundary condition. As remarked in Lecture 4 the estimates for $p = 2$ are special cases of more general results. We wish also to draw attention to a paper [22] by Hörmander concerned with equations $Lu = 0$ with constant coefficients in a half space, and solutions satisfying a number of boundary conditions $B_j u = 0$ described by differential operators $B_j$ with constant coefficients. Hörmander characterizes all such systems for which the solutions belong to $C^\infty$ on the boundary, and also those for which the solutions are analytic at the boundary.

Since the time is limited we shall restrict ourselves here to the Dirichlet problem for a single elliptic equations of order $2m$

$$(7.1) \qquad Lu = f \qquad \text{in } \mathcal{D}$$

$$\left(\frac{\partial}{\partial \vec{n}}\right)^{j-1} u = \varphi_s \qquad \text{on } \dot{\mathcal{D}}, j = 1, \ldots, m,$$

where $\vec{n}$ represents the unit normal to the boundary.

The operator $L$ will be required to satisfy a certain condition.

*Condition on $L$: If $L'(x, D)$ is the leading part of $L$, we require that for every pair of independent real vectors $\xi^1, \xi^2$ the polynomial in $\tau$*

$$L'(x, \xi^1 + \tau \xi^2)$$

*have exactly $m$ roots on either side of the real $\tau$ axis.*

In three or more dimensions this condition is automatically satisfied. We see however that the operator $\left(\frac{\partial}{\partial x} + i \frac{\partial}{\partial y}\right)^2$ in two dimensions violates the condition. (This operator and others in two dimensions come under the theory when treated as a system).

In describing the estimates we make use of the following norms and seminorms. In lecture 2 we already met the Hölder norm, for $0 < \alpha < 1$

$$[u]_\alpha = [u]_\alpha^{\mathcal{D}} = \underset{P, Q \in \mathcal{D}}{\text{l. u. b.}} \frac{|u(P) - u(Q)|}{|P - Q|^\alpha} .$$

For functions of classe $C^k$ in $\overline{\mathcal{D}}$ we also introduce (differing from the notation in Lecture 2)

$$[u]_k = \text{l. u. b.} \, |D^k u|$$

$$|u|_k = \sum_{j=0}^{k} [u]_j$$

where the l. u. b. is taken over all derivatives of order $k$, and all points in $\mathcal{D}$. In addition, for functions in $C^k$ with Hölder oontinuous (exponent $\alpha$) derivatives of order $k$, we introduce

$$[u]_{k+\alpha} = \max \, [D^k u]_\alpha ,$$

$$|u|_{k+\alpha} = |u|_k + [u]_{k+\alpha} ,$$

where the max, is taken over all derivatives of order $k$. The space of functions with finite $| \quad |_{k+\alpha}$ norm is denoted by $C^{k+\alpha}(\overline{\mathcal{D}})$. (We also use the notation of Lecture 3 and 4).

For functions $\varphi$ defined on the (smooth) boundary $\dot{\mathcal{D}}$ of $\mathcal{D}$ we also have analogous norms, defined in a tather obvious way en terms of local coordinates, and which we denote in the same way.

We now summarize the results without specifying the exact smoothness conditions on the boundary; $k$ will denote a non-negative integer. The integral estimates will be stated only for $p = 2$.

$L_2$ *Estimates: If* $u \in H_{2m}$ *and satisfies, for simplicity, homogeneous Dirichlet data,* i. e. $\varphi_j = 0$, *and if* $L u \in H_k$, *and the coefficients of* $L$ *belong to* $C^k$, *then* $u$ *belongs to* $H_{2m+k}$ *and*

$$\| u \|_{2m+k} \leq \text{constant} \; ( \| L u \|_k + \| u \|_0 ) . \tag{7.2}$$

*Similar results hold for equations in integral, or variational, form.*

*Schauder Estimates: If for some positive* $\alpha < 1$, $u \in C^{2m+\alpha}(\overline{\mathcal{D}})$, $Lu \in C^{k+\alpha}(\overline{\mathcal{D}})$, $\varphi_j \in C^{2m+k+1-j+\alpha}$, *and the coefficients of* $L$ *belong to* $C^{k+\alpha}$, *then* $u \in C^{2m+k+\alpha}$ *and*

$$| u |_{2m+k+\alpha} \leq \text{constant} \; ( [L u]_{k+\alpha} + \sum_j | \varphi_j |_{2m+k+1-j+\alpha} + | u |_0 ) . \tag{7.3}$$

*Similar results hold for equations in variational form. From these one may derive, dor instance, the following result for solutions of* $L u = 0$, *under suitable smoothness assumptions on the coafficients:*

*If* $u \in C^{m-1+\alpha}(\overline{\mathcal{D}})$ *and* $\varphi_j \in C^{m-j+k+\alpha}$, $j = 1, \dots, m$, *then* $u \in C^{m-1+k+\alpha}(\overline{\mathcal{D}})$ *and*

$$| u |_{m-1+k+\alpha} \leq \text{constant} \; ( \Sigma \, | \varphi_j |_{m-j+k+\alpha} + | u |_0 ) . \tag{7.4}$$

The constants in the above are independent of $u$. In case of uniqueness of the solution in the class considered the terms $\| u \|_0$ or $| u |_0$ may be dropped. Miranda [23], using results of Agmon [25], has recently proved an extended maximum principle for solutions of strongly elliptic equations (7.1) in two dimensions, which asserts that (7.4) holds for $k = \alpha = 0$. I believe that this holds true in general for operators satisfying the condition on $L$.

The Schauder estimates have a number of useful consequences. With their aid one may prove the existence of solutions of strongly elliptic equations having merely Hölder continuous coefficients. In particular, with the aid of (7.4) one may solve such equations with the given $\Phi_j$ in class $C^{m-j+\alpha}$.

In addition one can also solve the Dirichlet problem for a wide class of equations which are not strongly elliptic. Futhermore, and this is perhaps the most useful feature of the estimates, with their aid one may prove local parturbation theorems for nonlinear elliptic squations. For example if $F_\lambda(x, u, \dots, D^{2m} u) = 0$ is a nonlinear equation depending on a parameter $\lambda$ and «smoothly» on all variables, such that for $\lambda = 0$ the function $u_0$ is a solution with, say, zero Dirichlet data, and if the « first variation » of $F$ at $u_0$ is a linear elliptic operator $L$ which is invertible (i. e. for which the Dirichlet problem (7.1) has one and only one solution) then for $|\lambda|$ sufficiently small there exists a unique solution $u_\lambda$ of the nonlinear equation with zero Dirichlet data. The estimates also yield differentiability theorems at the boundary for solutions of nonlinear elliptic equations.

The estimates are derived following the « recipe » of Lecture 4, the main step being the first one. That is, one considers equations (7.1) in a half space $x_n > 0$, for operators $L$ with constant coefficients and only highest order terms, and $C^\infty$ functions $u$ in $t \geq 0$ vanishing outside some sphere. This system is then treated with the aid of explicitly constructed Poisson kernels, which will be described in the next lecture, with which one solves the system (7.1) with $f = 0$. With the aid of the explicit representations for $u$ and its derivatives so obtained, the desired estimates for this constant coefficient case are then obtained with the aid of certain potential theoretic results.

I would like to describe these results, which $I$ believe should prove useful for other problems. Since we are operating in a half space $x_n > 0$ it is convenient to rename the coordinates, set $(x_1, \dots, x_{n-1}) = x$, $x_n = t$, $(x_1, \dots, x_n) = (x, t) = P$.

We consider integral transforms of functions $f(x)$ into functions $u(x, t)$, $t>0$. Let $K(x,t)$ be a kernel defined in the half space $t \geq 0$ and homogeneous of degree $1 - n$.

$$K(P) = \frac{\Omega(x/|P|, t/|P|)}{|P|^{n-1}}$$

here $|P| = (|x|^2 + t^2)^{1/2}$; Assume that $\Omega$ is continuous on the half sphere $|P| = 1$, $t \geq 0$ and assume also (this condition can be weakened considerably) that $\Omega$ has continuous first derivatives on the half sphere which, together with $\Omega$ itself, are bounded in absolute value by $\varkappa$. In addition we make the basic assumption

$$\int_{|x|=1} \Omega(x, 0)\, d\omega_x = 0 .$$

Here integration is over the unit sphere $|x| = 1$, with $d\omega_x$ as element of area.

Consider the transformation

$$u(x, t) = \int K(x - y, t) f(y)\, dy \qquad t > 0,$$

integration being over the entire $y$ space. Denote the $L_p$ norm of $f$ by $|f|_{L_p}$, and that of $u(x, t)$ in $x$, for any fixed $t$, by $|u|_{L_{p,t}}$.

THEOREM: 1. *For* $0 < \alpha < 1$

$$[u]_\alpha \leq c \cdot \varkappa\, [f]_\alpha$$

*where* $c$ *depends only on* $\alpha$ *and* $n$. *Here the norms refer to the half space* $t > 0$ *for* $u$, *and the plane* $t = 0$ *for* $f$.

2. *For every* $t > 0$ *and* $1 < p < \infty$,

$$|u|_{L_{p,t}} \leq c\, \varkappa\, |f|_{L_p}$$

*where* $c$ *depends only on* $p$ *and* $n$.

3.
$$\left[\iint |u(x, t)|^2 \, dx \, dt\right]^{1/2} \leq c\, x\, \langle f \rangle_{-1/2}$$

*where* $c$ *is an absolute constant. Here* $\langle f \rangle_{-1/2}$ *is defined in terms of the Fourier transform* $\widetilde{f}(\xi)$ *of* $f$ *by*

$$\langle f \rangle_{-1/2} = \int |\xi|^{-1} |\widehat{f}(\xi)|^2 \, d\xi]^{1/2}.$$

There is an $L_p$ analogue of 3, which is however more complicated to state.

We call Part 1 of the theorem a result of Privaloff type. It is a simple extension of classical results of Hölder, Giraud and others, to which it reduces if we set $t = 0$. Part 2, a result of Riesz type, is a straighforwatd extension of recent results of Calderon and Zygmund [24], to which it reduces if we set $t = 0$. For the special case of the Hilbert transform for $n = 2$ it is due to Riesz, and in fact it is proved by reduction to the Riesz result with the aid of a device of [24]. Part 3, is proved with the aid of Fourier transforms — one shows that the Fourier transform $\widehat{K}(\xi, t)$ of $K(x, t)$ with respect to the $x$ variables is bounded in absolute value by constant $(1 + t|\xi|)^{-1}$, from which the result follows easily. Part 3 plays an essential role in the derivation of the $L_2$ estimates.

## Lecture VIII. The Boundary Value Problem in a Half Space; The Poisson Kernels.

In this lecture we shall show how to solve explicity the elliptic system (7.1) with constant coefficients for the special case of a half space. Making a slight change of notation we shall consider the space to be

$n+1$ dimensional, with the first $n$ coordinates denoted by $x=(x_1, \dots, x_n)$ and the last coordinate by $t$. In the half space $t>0$ we consider for simplicity the homogeneous equation, with $D_x = \left(\frac{\partial}{\partial x_1}, \dots, \frac{\partial}{\partial x_n}\right)$, $D_t = \frac{\partial}{\partial t}$,

$$L(D_x, D_t)\, u = 0 \tag{8.1}$$

where $L$ is an elliptic operator of order $2m$ with only highest order terms, satisfying the « condition on $L$ » of the previous lecture, i. e. for fixed real $\xi = (\xi_1, \dots, \xi_n) \neq 0$ the polynomial $L(\xi, \tau)$ has exactly $m$ roots $\tau$ on each side of the real axis.

On $t=0$ we prescribe the derivatives

$$D_t^{j-1} u = \Phi_j(x) \qquad j = 1, \dots, m \tag{8.2}$$

with the $\Phi_j$ in $C_0^\infty$, for simplicity.

The solution will be given in terms of kernels $K_j(x, t)$, $j = 1, \dots, m$, the Poisson kernels,

$$u(x, t) = \Sigma_j \int K_j(x-y, t)\, \Phi_j(y)\, dy = \Sigma\, K_j * \Phi_j, \tag{8.3}$$

where $*$ denotes convolution. Our construction of the $K_j$ is an extension of the construction given by Agmon [25] in two dimensions, $n=1$, but it is based on the Fritz John identity (1.6) of Lecture 1: For $\Phi(x)$ in $C_0^\infty$

$$\Phi = -\frac{1}{(2\pi i)^n q!}\, \Delta^{(n+q)/2} \left[ \int\limits_{|\xi|=1} (x\cdot\xi)^q \log \frac{x\cdot\xi}{i}\, d\omega_\xi * u \right], \tag{8.4}$$

where $q$ is a non-negative integer of the same parity as $n$, $\Delta$ is the Laplacean, and the principal branch of the lagarithm is taken with the plane slit along the negative real exis.

First some preliminaries. For fixed real $\xi \neq 0$ denote by $\tau_k^+ = \tau_k^+(\xi)$, $k = 1, \dots, m$, the roots $\tau$ with positive imaginary parts of $L(\xi, \tau) = 0$, and set

$$M^+(\xi, \tau) = \prod_k (\tau - \tau_k^+(\xi)) = \sum_{p=0}^{m} a_p^+(\xi)\, \tau^{m-p}.$$

The coefficients $a_p^+$ are analytic in $\xi$ for real $\xi \neq 0$, and homogeneous of degree $p$. With $M^+$ we associate the polynomials (in $\tau$)

$$M_j^+(\xi, \tau) = \sum_{p=0}^{j-1} a_p^+(\xi)\, \tau^{j-1-p}, \qquad j = 1, \dots, m. \tag{8.5}$$

The following relations are easily verified.

$$\frac{1}{2\pi i}\int_{\gamma} \frac{M^{+}_{m+1-j}(\xi,\tau)}{M^{+}(\xi,\tau)}\tau^{k-1}\,d\tau=\delta^{k}_{j}, \qquad 1\le j,\ k\le m \tag{8.6}$$

where $\gamma$ is a rectifiable Jordan contour in the complex $\tau$ plane enclosing all the roots $\tau^{+}(\xi)$ in its interior; $\delta^{k}_{j}$ is the Kronecker delta.

We can now writhe down the

*Poisson Kernels*: For $j-1\ge n$

$$K_j(x,t)=\frac{\beta_j}{2\pi i}\int_{|\xi|=1} d\omega_\xi\left[\int_{\gamma}\frac{M^{+}_{m+1-j}(\xi,\tau)}{M^{+}(\xi,\tau)}(x\cdot\xi+t\tau)^{j-1-n}\log\frac{x\cdot\xi+t\tau}{i}\,d\tau\right], \tag{8.7}$$

for $j-1<n$

$$K_j(x,t)=\frac{\beta_j}{2\pi i}\int_{|\xi|=1} d\omega_\xi\left[\int\frac{M^{+}_{m+1-j}(\xi,\tau)}{M^{+}(\xi,\tau)(x\cdot\xi+t\tau)^{n-j+1}}\,d\tau\right], \qquad j=1,\dots,m. \tag{8.7$'$}$$

Here

$$\begin{aligned}\beta_j&=-\frac{1}{(2\pi i)^n(j-1-n)!} &&\text{if } j-1\ge n,\\ \beta_j&=(-1)^{n-j+1}\frac{(n-j)!}{(2\pi i)^n} &&\text{if } j-1<n,\end{aligned} \tag{8.8}$$

and $\gamma$ is a Jordan contour in $\mathfrak{Im}\,\tau>0$ enclosing all the roots $\tau$ of $M^{+}(\xi,\tau)$ *for all* $|\xi|=1$, $\xi$ real.

Before proving that these formulas represent Poisson kernels we observe, with the aid of the identities

$$\begin{aligned}&\frac{\mu!}{(\lambda+\mu)!}\left(\frac{d}{dz}\right)^{\lambda}\left[z^{\lambda+\mu}\left(\log\frac{z}{i}+c_{\lambda,\mu}\right)\right]=z^{\mu}\log\frac{z}{i}, && \mu,\lambda\ge 0,\\ &\frac{(-1)^{1+\mu}}{(\mu+\lambda)!\,(-1-\mu)!}\left(\frac{d}{dz}\right)^{\lambda}\left(z^{\lambda+\mu}\log\frac{z}{i}\right)=z^{\mu}, && \mu<0,\ \lambda+\mu\ge 0\end{aligned} \tag{8.9}$$

for $\lambda$, $u$ integers, and $c_{\lambda,\mu}$ some appropriate constants, that we may represent the functions $K_j$ in the form — with $q$ a non-negative integer having the same parity as $n$ —

$$K_j=\Delta_x^{(n+q)/2}K_{j,q}(x,t) \tag{8.10}$$

where, for $j-1\geq n$,

$$K_{j,q}=\frac{\beta_j\,(j-1-n)\,!}{2\pi i\,(j-1+q)\,!}\cdot \tag{8.10$'$}$$

$$\cdot\int\limits_{|\xi|=1} d\omega_\xi\left[\int\limits_{\gamma}\frac{M^+_{m+1-j}\,(x\cdot\xi+t\,\tau)^{j-1+q}}{M^+}\left(\log\frac{x\cdot\xi+t\,\tau}{i}+c_{n+q,\,j-1-n}\right)d\tau\right],$$

and for $j-1<n$

$$K_{j,q}=\frac{(-1)^{n-j}}{2\pi i\,(j-1+q)\,!\,(n-j)\,!}\cdot \tag{8.10$''$}$$

$$\cdot\int\limits_{|\xi|=1} d\omega_\xi\left[\int\limits_{\gamma}\frac{M^+_{m+1-j}\,(x\cdot\xi+t\,\tau)^{j-1+q}}{M^+}\log\frac{x\cdot\xi+t\,\tau}{i}\,d\tau\right].$$

It is easily seen that $K_{j,q}$ and all its derivatives up to order $j+q$ are continuous in the closed half space $t\geq 0$.

We now prove that the kernels $K_j$ given by (8.7), (8.7)$'$ are indeed Poisson kernels. By inspection we see that the $K_j$ are analytic solutions of $Lu=0$ for $t>0$. Hence $u$ defined by (8.3) is a solution. Setting

$$u_j=K_j * \Phi_j. \tag{8.3$_j$}$$

we shall show that $u_j$ belongs to $C^\infty$ in $t\geq 0$ and that

$$D_t^{k-1}\,u_j=\delta_j^k\,\Phi_j(x)\qquad\text{for } t=0,\ k=1,\dots,m. \tag{8.11}$$

Consider any partial derivative of order $s$ of $u_j$. Choosing an integer $q$ of the same parity as $n$, and such that $q\geq s-j+2$ we have, for $t>0$

$$D^s\,u_j=D^s\int \Delta_x^{(n+q)/2}\,K_{j,q}\,(x-y\,,\,t)\,\Phi_j\,(y)\,dy \tag{8.12}$$

$$=\int D^s\,K_{j,q}\,(x-y\,,\,t)\,\Delta_y^{(n+q)/2}\,\Phi_j\,(y)\,dy$$

after partial integration, recalling that $\Phi_j\in C_0^\infty$. Since, as remarked above, $D^s K_{j,q}$ is continuous in the closed half space $t\geq 0$ it follows that $D^s u_j$ can be extended as a continuous function in the entire closed half space $t\geq 0$. Since $s$ is arbitrary we have proved that $u_j\in C^\infty$ in $t\geq 0$.

To verify (8.11) choose $q$ sufficiently large so that $q \geq j - k + 1$, $j = 1, \ldots, m$. Using (8.12) we have, for $t = 0$,

$$D_t^{k-1} u_j = \int \Delta_y^{(n+q)/2} \Phi_j(y) \cdot D_t^{k-1} K_{j,q}(x-y, 0)\, dy \tag{8.13}$$

$$= \int \Delta_x^{(n+q)/2} \Phi_j(x-y)\, D_t^{k-1} K_{j,q}(y, 0)\, dy ,$$

after a change of variable.

Assume first that $k \neq j$. Using (8.10)′, (8.10)″ we find, for $t = 0$, and appropriate constants $c'$, $c''$

$$D_t^{k-1} K_{j,q}(y, 0) = c' \int_{|\xi|=1} \int_\gamma \frac{M^+_{m+1-j}}{M^+} \tau^{k-1} d\tau\, (y \cdot \xi)^{j-k+q} \left(\log \frac{y \cdot \xi}{i} + c''\right) d\omega_\xi = 0$$

by (8.6). Thus (8.11) is proved for $k \neq j$.

Now suppose $k = j$. If $j - 1 > n$ we have, using (8.9) (8.10)′ and (8.6), for some constant $c'$

$$D_t^{j-1} K_{j,q}(y, 0) = \tag{8.14'}$$

$$= \frac{\beta_j (j-1-n)!}{2\pi i\, q!} \int_{|\xi|=1} d\omega_\xi \left[ (y \cdot \xi)^q \left(\log \frac{y \cdot \xi}{i} + c'\right) \int_\gamma \frac{M^+_{m+1-j}}{M^+} \tau^{j-1} d\tau \right]$$

$$= \frac{\beta_j (j-1-n)!}{q!} \int_{|\xi|=1} (y \cdot \xi)^q \log \frac{y \cdot \xi}{i} + \psi_q(y)$$

where $\psi_q(y)$ is a homogeneous polynomial of degree $q$.

Similarly if $j - 1 < n$ we find, using (8.10)″ and (8.6)

$$D_t^{j-1} K_{j,q}(y, 0) = \frac{(-1)^{n-j} \beta_j}{(n-j)!\, q!} \int_{|\xi|=1} (y \cdot \xi)^q \log \frac{y \cdot \xi}{i}\, d\omega_\xi + \psi_q(y) \tag{8.14''}$$

where again $\psi_q$ denotes a homogeneous polynomial of degree $q$.

From (8.13), (8.14)′, (8.14)″ we find, after inserting the value of $\beta_j$ from (8.8), and rechanging variables, that

$$D_t^{j-1} u_j(x, 0) = -\frac{1}{(2\pi i)^n q!} \Delta_x^{(n+q)/2} \int \Phi_j(y) \int_{|\xi|=1} ((x-y) \cdot \xi)^q \cdot \tag{8.15}$$

$$\cdot \log \frac{(x-y) \cdot \xi}{i}\, d\omega_\xi\, dy .$$

Here we have used the fact that

$$\int \Delta_y^{(n+q)/2} \Phi_j(y) \cdot \psi_q(x-y)\, dy = \int \Phi_j(y) \cdot \Delta_x^{(n+q)/2} \psi_q(x-y)\, dy = 0$$

since $\psi_q$ is a polynomial of degree $q$ and is therefore annihilated by $\Delta_x^{(n+q)/2}$. By John's identity (8.4) the right side of (8.15) equals $\Phi_j(x)$, and the proof that the $K_j$ are Poisson kernels is complete.

We remark that the functions $K_{j,q}$ are actually analytic in $t \geq 0$ except at the origin, and that for $s \geq j+q$, $D^s K_{j,q}$ is homogeneous of degree $j-1+q-s$. It follows from our proof above that if $s = n+q+j-k \geq 0$ then

$$(8.16) \qquad | D_x^s D_t^{k-1} K_{j,q}(x, t) | \leq \text{constant} \cdot \frac{t}{(|x|^2 + t^2)^{(n+1)/2}}, \qquad k \neq j.$$

Furhermore we see that because of the reproducing properties (8.11) of the $K_j$ we may assert that

$$D_t^{j-1} K_j(x, 0) = 0 \qquad \text{for } x \neq 0,$$

or

$$(8.16)' \qquad D_t^{j-1} K_j(x, t) \leq \text{constant} \frac{t}{(|x|^2 + t^2)^{(n+1)/2}}.$$

With the aid of (8.16), (8.16)′ it is not difficult to establish the following

*Extended Maximum Principle: The solution* (8.3) *of the Dirichlet problem* (8.1), (8.2) *satisfies*

$$\text{l. u. b.} \, | D^{m-1} u(x, t) \leq \text{constant l. u. b.} \, | D^{m-1} u(x, 0) |$$

*where the least upper bounds are taken with respect to all derivatives of order* $m-1$ *and, on the left, with respect to all* $(x, t)$ *in the half space, on the right with respect to all* $x$.

This is an analogue of a special case of Miranda's extended maximum principle of [23].

# BIBLIOGRAPHY

[1] H. LEWY, *An example of a smooth linear partial differential equation without solution.* Annals of Math. 66 (1957) p. 155-158.

[2] L. EHRENPREIS, *Solutions of some problems of division I, II.* American Journal of Math 76 (1954) p. 883-903, 77 (1955) p. 286-292. *The division problem for distributions.* Proc. Nat. Acad. Sci. 41-10 (1955) p. 756-758.

[3] L. HÖRMANDER, *On the theory of general partial differential operators.* Acta Math. 94 (1955) p. 160-248.

[4] B. MALGRANGE, *Existence et approximation des solutions des équations aux dérivées partielles et des équations de convolutions.* Annales de L'Inst. Fourier 6 (1955-6) p. 271-355.

[5] F. TRÈVES, *Solutions élémentaire d'équations aux dérivées partielles dépendent d'un paramètre,* C. R. Acad. Sci. Paris 242 (1956) p. 1250-1252.

[6] L. HÖRMANDER, *Local and global properties of fundamental solutions.* Math. Scandinavica 5 (1957) p. 27-39. *On the division of distributions by polynomials.* Arkiv. för Mat. 3 No. 53 (1958) p. 555-568.

[7] F. JOHN, *Plane waves and spherical means applied to partial differential equations.* Interscience, New York, 1955.

[8] N. DU PLESSIS, *Some theorema about the Riesz fractional integral.* Trans. Amer. Math. Soc. 80 (1955) p. 124-134.

[9] L. NIRENBERG, *Remarks on strongly elliptic partial differential equations.* Comm. Pure Appl. Math. 8 (1955) p. 649-675.

[10] L. BERS, L. NIRENBERG, (a) *On a representation theorem for linear elliptic systems with discontinuous coefficients and its applications.* (b) *On linear and nonlinear elliptic boundary value problems in the plane.* Atti de Convegno Inter. sul. Equazioni alle derivate Parziali, Trieste, 1954 (published 1955).

[11] DE GIORGI, *Sulla differenziabilità e l'analiticità delle estremali degli integrali multipli regolari.* Mem. della Accad. delle Scienze di Torino. Ser. 3, Vol. 3 (1957) p. 25-43.

[12] J. NASH, *Cantinuity of solutions of parabolic and elliptic equations.* Amer. Journ. Math. 80 (1958) p. 931-954

[13] N. ARONSZAJN, *On coercive integro differential quadratic forms.* Conference on Partial Differential Equations, University of Kansas, 1954, Technical Report No. 14, p. 94-106.

[14] S. AGMON, A. DOUGLIS, L. NIRENBERG *Estimates near the boundary for sotutions of elliptic partial differential equations satisfying general boundary conditions I.* To appear in Comm. Pure Appl. Math.

[15] M. SCHECHTER, *Integral inequalities for partial differential operators and functions satisfying general boundary conditions.* To appear in Comm. Pure Appl. Math. Vol. 12, No. 1 (1959).

[16] S. AGMON, *The coerciveness problem for integro differential forms,* Journal d'Analyse Math. 6 (1958) p. 184-223.

[17] L. HÖRMANDER, *On the interior regularity of the solutions of partial differential equations.* Comm. Pure Appl. Math. 11 (1958) p. 197-218.

[18] B. MALGRANGE, *Sur une classe d'opérateurs différentiels hypoelliptiques.* Bull. Soc. Math. France. 85, 3 (1957) p. 283-306.

[19] P. D. LAX, *On Cauchy's problem for hyperbolic equations and the differentiability of solutions of elliptic equations*. Comm. Pure Appl. Math. 8 (1955) p. 615 633.

[20] L. BERS, *Elliptic partial differential equations*. Lecture notes of the Seminar in Applied Mathematics, University of Colorado, June 1957.

[21] L. SCHWARTZ, *Ecuaciones diferenciales parciales elipticas*, Lectures at Bogota Colombia, 1956.

[22] L. HÖRMANDER, *On the regularity of the solutions of boundary problems*. Acta Math 99 (1958) p. 225-264.

[23] C. MIRANDA, *Teorema del massimo modulo e teorema di esistenza e di unicità per il problema di Dirichlet relativo alle equazioni ellittiche in due variabili*. Annali di Mat. Pura ed Appl. Ser. 4, Vol. 46 (1958) p. 265-312.

[24] A. P. CALDERON, A. ZYGMUND, *On singular integrals*, Amer. Journal Math. 79 (1956) p. 289-309.

[25] S. AGMON, *Multiple layer potentials and the Dirichlet problem for higher order elliptic equations in the plane I*. Comm. Pure Appl. Math. 10 (1957) p. 179-239.

## ADDITIONAL GENERAL BIBLIOGRAPHY

[A] C. MIRANDA, *Equazioni alle derivate parziali di tipo ellittico*, Springer, Berlin 1955.

[B] *Transactions of the Symposium on Partial Differential Equations*. Berkeley California, 1955, published in Comm. Pure Appl. Math., Vol. 9, No. 3 (1956).

[C] E. MAGENES, G. STAMPACCHIA, *I problemi al contorno per le equazioni differenziali lineari di tipo ellittico*. Annali della Scuola Norm. Sup. di Pisa Ser. 3, Vol. 12, Fasc. 3 (1958) p. 297-358.

Estratto dagli *Annali della Scuola Normale Superiore di Pisa*
Serie III. Vol. XIII. Fasc. IV (1959)

# THE $L_p$ APPROACH TO THE DIRICHLET PROBLEM (*)

by SHMUEL AGMON

## PART I
## REGULARITY THEOREMS

### 1. Introduction.

In this paper we present a $L_p$ approach to the Dirichlet problem and to related regularity problems for higher order elliptic equations. Although this approach is not as simple as the well known Hilbert space approach developed by Vishik [32] Gårding [14], Browder [6 ; 7], Friedrichs [12], Morrey [22], Nirenberg [23], Lions [18] and others, it has the advantage of a greater generality. Thus, for example, we shall be able to treat the non-homogeneous Dirichlet problem in a much more general situation not restricted to solutions having a finite Dirichlet integral (in this connection see Magenes-Stampacchia [19, § 9] and the recent paper of Miranda [20]). The method is also applicable to elliptic operators which are not necessarily strongly elliptic. We remark further that the same method could be used to solve a general class of boundary value problems. This will be done in a subsequent paper where we shall also derive $L_p$ integral inequalities for a system of differential operators acting on functions satisfying general boundary conditions, simular to the « coercive » $L_2$ inequalities derived by Aronszajn [4] Agmon [2] and Schechter [25].

Recently Schechter [26 ; 27] presented a Hilbert space approach to general boundary value problems including the Dirichlet problem for non-strongly elliptic equations. His method is based on the $L_2$ estimates of Agmon-

---

(*) Presented in part (for $p = 2$) at the international conference on partial differential equations organized by the C. I. M. E. in Pisa, September 1-10, 1958. Sponsored in part by the Office of Scientific Research of the A. R. D. C., U. S. Air Force, through its European Office, under Contract No. AF 61 (052)-187.

Douglis-Nirenberg [3] (see also [2; 25]) and on known $L_2$ regularity theorems. Our $L_p$ method which utilizes new regularity theorems is quite different and the results we obtain are stronger in various respects (1). Other existence results utilizing the continuity method were given by Agmon-Douglis-Nirenberg [3].

The $L_p$ approach to the Dirichlet problem is based on a $L_p$ regularity theory for very weak solutions of the Dirichlet problem. To obtain such a regularity theory we use some of the ideas of a method originally devised by Nirenberg [23] with the following essential modification: instead of using Gårding's inequality we use the explicit solution of the Dirichlet problem for elliptic operators with constant coefficients in a half-space, and the $L_p$ estimates for such solutions derived in [3].

The paper is divided into two parts. In Part I we give the basic regularity theory, both in the interior and at the boundary. This part has an independent interest and entails most of the work. We remark that when we consider the simpler problem of interior regularity we consider also weak solutions of overdetermined elliptic systems and derive $L_p$ estimates for such solutions. In Part II we shall combine the regularity theory with some general results on Banach spaces (using in particular a result of Fichera [10]) to develop the $L_p$ existence theory for the Dirichlet problem.

## 2. Notations and definitions.

Throughout the paper we denote by $G$ a bounded domain in $n$ dimensional space with boundary $\partial G$ and closure $\bar{G}$. We denote by $x = (x_1, \dots, x_n)$ the generic point in the space and put $|x| = (x_1^2 + \dots + x_n^2)^{\frac{1}{2}}$. We say that $G$ is of class $C^j$ if with every point $x^0 = (x_1^0, \dots, x_n^0) \in \partial G$ one can associate a sphere $S$ having its center in $x^0$ such that $\partial G \cap S$ admits a representation of the form:

$$x_k = g(x_1, \dots, x_{k-1}, x_{k+1}, \dots, x_n) \tag{2.1}$$

for a suitable $k$; $g$ being a function defined in some neighborhood $\mathsf{U}$ of $(x_1^0, \dots, x_{k-1}^0, x_{k+1}^0, \dots, x_n^0)$ possessing there continuous derivatives up to the order $j$.

---

(1) For instance, Schechter's method is applicable only to such problems for which the solution of the adjoint problem is unique, whereas we get the alternative in the general case without any uniqueness assumption.

Similarly, $G$ is said to be of classe $C^{0,1}$ if around each of its points the boundary $\partial G$ admits a local representation of the form (2.1) with a function $g$ satisfying a Lipschitz Condition in the neighborhood $\mathsf{U}$.

Finally, $G$ is said to possess the cone property if every point in $\overline{G}$ is a vertex of a closed right spherical cone of fixed opening and height which belongs to $\overline{G}$. It is readily seen that if $G$ is of classe $C^{0,1}$ then it also has the cone property.

We shall denote by $C^k(G)$ (resp. $C^k(\overline{G})$) the class of complex valued functions possessing continuous derivatives up to ther order $k$ $(0 \leq k \leq \infty)$ in $G$ (resp. $\overline{G}$). The class of infinitely differntiable functions with compact support in $G$ will be denoted by $C_0^\infty(G)$.

Let, now, $j$ be a non-negative integer and $p$ a real number $\geq 1$. For a function $u(x)$ belonging to $C^j(\overline{G})$ define the norm:

$$\| u \|_{j,L_p(G)} = \left( \int_G \sum_{|\alpha| \leq j} | D^\alpha u |^p \, dx \right)^{1/p}, \tag{2.2}$$

where here and in the following $\alpha$ stands for the multi-index $(\alpha_1, \dots, \alpha_n)$, $|\alpha| = \alpha_1 + \dots + \alpha_n$, and $D^\alpha$ is the partial derivative:

$$D^\alpha = \frac{\partial^{\alpha_1 + \dots + \alpha_n}}{\partial x_1^{\alpha_1} \dots \partial x_n^{\alpha_n}}.$$

We also put:

$$D_i = \frac{\partial}{\partial x_i} \quad (i = 1, \dots, n) \quad \text{and} \quad D = (D_1, \dots, D_n).$$

The linear space $C^j(\overline{G})$ is clearly not complete under the norm (2.2). Completing it we obtain a Banach space which we denote by $H_{j,L_p}(G)$. We retain the notation $\| \quad \|_{j,L_p(G)}$ for the norm in $H_{j,L_p}(G)$. The space $H_{0,L_p}(G)$ is simply the space $L_p(G)$ and we shall usually write $\| \quad \|_{L_p(G)}$ for the norm in this space.

The classes of functions $H_{j,L_p}(G)$ were investigated by many authors (Sobolev [30], Morrey [21], Friedrichs [11], Stampacchia [31], Deny and Lions [9], Gagliardo [13] and others). Some of the properties of these classes will be described in the next section. Here we limit ourselves to some remarks.

By the identification mapping we can consider $H_{j,L_p}(G)$ as a linear subset of $L_p(G)$. A function $u \in H_{j,L_p}(G)$ will possess generalized derivatives up to the order $j$ which we term strong $L_p$ derivatives. To define these let

$\{u_i\}_{i=1}^{\infty}$ be a sequence of functions in, $C^j(\overline{G})$ such that

$$\lim_{i\to\infty} \| u_i - u \|_{j,L_p(G)} = 0 .$$

Then, there exist functions $u^\alpha \in L_p(G)$ $(0 \leq |\alpha| \leq j)$ such that

$$\lim_{i\to\infty} \| D^\alpha u_i - u^\alpha \|_{L_p(G)} = 0 .$$

The functions $u^\alpha$ are by definition the strong $L_p$ derivatives $D^\alpha u$ of $u$ in $G$. They are uniquely defined.

We shall say that a function $u$ belong locally to $H_{j,L_p}$ in $G$ — writing $u \in H_{j,L_p}^{\text{loc.}}(G)$ — if for every $x \in G$ there exists a sphere $S \subset G$ with center at $x$ such that $u \in H_{j,L_p}(S)$. It is readily seen (using a partition of unity) that if $u \in H_{j,L_p}^{\text{loc.}}(G)$ then $u \in H_{j,L_p}(G_1)$ for every domain $G_1$ such that $\overline{G}_1 \subset G$.

In connection with the Dirichlet problem we shall have to consider the subclass of functions in $H_{j,L_p}$ which together with some of their derivatives, vanish at the boundary in a generalized sense. To make this more precise suppose that $G$ is of class $C^{0,1}$. Let $u \in H_{1,L_p}(G)$. Then, as it is well known, one can define for such $u$ its trace $\gamma(u)$ on the boundary. For instance, one can use the following procedure. For $u \in C^1(\overline{G})$ $\gamma(u)$ is simply the restriction of $u$ on $\partial G$. In this case it is easily established that

$$\left( \int_{\partial G} |\gamma(u)|^p \, d\sigma \right)^{1/p} \leq \text{Const.} \| u \|_{1,L_p}(G) \tag{2.3}$$

with a constant which is independent of $u$. Hence, $u \to \gamma(u)$ is a bounded linear transformation from $C^1(\overline{G})$ (considered as a subset in $H_{1,L_p}(G)$) into $L_p(\partial G)$. Since $C^1(\overline{G})$ is dense in $H_{1,L_p}(G)$ one can extend the transformation in a unique manner by continuity to the whole of $H_{1,L_p}(G)$. This defines the trace on the boundary of a function $u \in H_{1,L_p}(G)$ as an element of $L_p(\partial G)$.

Let, now, $m, j$ be positive integers such that $m \leq j$. We denote by $H_{j,L_p}(G ; \{D^\alpha\}_{|\alpha| \leq m-1})$ the class of functions $u \in H_{j,L_p}(G)$ which satisfy the boundary conditions

$$D^\alpha u = 0 \quad \text{on} \quad \partial G \quad \text{for} \quad 0 \leq |\alpha| \leq m-1 , \tag{2.4}$$

where (2.4) is taken in the sense that

$$\gamma(D^\alpha u) = 0 \quad \text{as an element of } L_p(\partial G).$$

We observe that $H_{j,L_p}(G\,;\,\{D^\alpha\}_{|\alpha|\leq m-1})$ is a closed subspace of $H_{j,L_p}(G)$, and that a function $u$ belonging to $H_{j,L_p}(G\,;\,\{D^\alpha\}_{|\alpha|\leq m-1}) \cap C^{m-1}(\bar{G})$ satisfies the boundary conditions (2.4) pointwise in the ordinary sense.

## 3. Calculus and properties of the classes $H_{j,L_p}$.

We have remarked already that a function belonging to $H_{j,L_p}(G)$ possesses strong $L_p$ derivates up to the order $j$ in $G$. Considering such a function $u$ as a distribution in $G$ (Schwartz [28]), it is readily seen that the strong $L_p$ derivatives are also the distribution derivatives of $u$ which are thus functions belonging to $L_p(G)$. It is very convenient that under general conditions on the domain $G$ one can reverse this statement. We have:

THEOREM 3.1. *Suppose that $G$ is of class $C^{0,1}$. Let $u \in L_p(G)\,(p \geq 1)$ and assume that the distribution derivatives of $u$ of order $\leq j$ are functions belonging to $L_p(G)$. I. e., assume that there exist functions $u^\alpha(x) \in L_p(G)$, $0 < |\alpha| \leq j$ (weak derivatives in the terminology of Friedrichs) such that*

$$\int_G u D^\alpha \varphi \, dx = (-1)^{|\alpha|} \int_G u^\alpha \varphi \, dx, \tag{3.1}$$

*for all $\varphi \in C_0^\infty(G)$. Then, $u \in H_{j,L_p}(G)$ and its distribution derivatives $u^\alpha$ coincide with its strong $L_p$ derivatives $D^\alpha u$ ($|\alpha| \leq j$).*

The weaker conclusion that $u \in H^{\text{loc.}}_{j,L_p}(G)$ is well known and was established by various authors (Friedrichs [11], Sobolev [30], Deny-Lions [9]). The theorem as stated is due to Gagliardo [2] [13]. For more regular domains it was established by Babich [5].

The following remarks are obvious. If $u \in H_{j,L_p}(G)$ and $a \in C^j(\bar{G})$, then $v = au$ belongs to $H_{j,L_p}(G)$ and the strong derivatives of $v$ are obtained by the standard Leibniz rule. If, moreover, $G$ is of class $C^{0,1}$ then the boundary values $\gamma(D^\alpha v)$ ($|\alpha| \leq j-1$) are obtained by the same rules. The classes $H_{j,L_p}$ are preserved by homeomorphism of class $C^j$. That is, let $x^* \to x(x^*)$ be a one to one mapping of $\bar{G}^*$ onto $\bar{G}$ such that the mapping and its inverse possess continuous derivatives up to the order $j$ in the corresponding closed domains. Then the mapping $u \to u^*$, $u^*(x^*) = u(x(x^*))$, is a homeo-

(2) It should be pointed out that Gagliardo is not using the notion of a weak derivative but a different notion which is, however, equivalent to it. Also, the proof of the main approximation theorem [13; p. 112] could be repeated word by word for functions possessing weak derivatives in the sense of (3.1).

morphism between $H_{j,L_p}(G)$ and $H_{j,L_p}(G^*)$. Also, to compute the strong derivatives of $u^*$ one applies the usual chain rule. The same remark applies to the trace at the boundary of derivatives of order $\leq j-1$ when $G$ is of class $C^{0,1}$.

Most of the following lemmas are the $L_p$ modified versions of the calculus $L_2$ lemmas given in Nirenberg [23]. Unless otherwise stated we shall assume in these lemmas that $G$ is of class $C^{0,1}$.

LEMMA 3.1. *Let $u \in L_p(G)$, $p > 1$. Suppose that $u$ is a weak limit in $L_p$ of a sequence of functions $\{u_k\}$ which belong to $H_{j,L_p}(G)$ and possess uniformly bounded norms $\| u_k \|_{j,L_p(G)}$. Then, $u \in H_{j,L_p}(G)$ and its derivatives of order $\leq j$ are the weak $L_p$ limits of the corresponding derivatives of the functions $u_k$.*

*Proof*: From the weak compactess of the unit sphere in $L_p\,(p > 1)$ it follows that there there exists a subsequence $u_{k'}$ such that $D^\alpha u_{k'}$ converges weakly in $L_p$ to a function $u^\alpha \in L_p(G)$, $|\alpha| \leq j$. Hence, for every function $\varphi \in C_0^\infty(G)$:

$$\int_G \varphi\, u^\alpha\, dx = \lim_{k'\to\infty} \int_G \varphi\, D^\alpha u_{k'}\, dx$$

$$= \lim_{k'\to\infty} (-1)^{|\alpha|} \int_G D^\alpha \varphi \cdot u_{k'}\, dx = (-1)^{|\alpha|} \int_G D^\alpha \varphi \cdot u\, dx .$$

Thus, $u^\alpha$ is the distribution (weak) derivative $D^\alpha u$. But then, since $u^\alpha \in L_p(G)$, it follows from Theorem 3.1 that $u \in H_{j,L_p}(G)$ and that $u^\alpha$ coincides with the strong $L_p$ derivative $D^\alpha u$. Moreover, from the uniqueness of the derivatives it follows that the whole sequence $D^\alpha u_k$ converges weakly to $D^\alpha u$ and not only a subsequence.

Using Theorem 3.1 one also obtains readily the following

LEMMA 3.2. *Suppose that $u$ belongs to $H_{j,L_p}(G)$ and that its $j'$ th order derivatives belong to $H_{k,L_p}(G)$, then $u \in H_{j+k,L_p}(G)$.*

NOTATION: Let $h = (h_1, \dots, h_n)$ be a real non-vanishing vector. We shall use the symbol $\delta_h$ to denote the difference quotient operator:

$$\delta_h u = \frac{u(x+h) - u(x)}{|h|} . \tag{3.2}$$

LEMMA 3.3. *Let $u \in H_{j,L_p}(G)$ $(j \geq 0, p > 1)$. Suppose that there exists a constant $C$ such that for every subdomain $G_1$; $\overline{G}_1 \subset G$:*

$$\| \delta_h u \|_{j,L_p(G_1)} \leq C \tag{3.3}$$

*for all sufficently small vectors h. Then* $u \in H_{j+1,L_p}(G)$ and

(3.4) $$\| D_i u \|_{j,L_p(G)} \leq C, \quad i = 1, \ldots, n.$$

*Proof:* Consider first the case $j = 0$. From (3.3) and the weak compactness of the unit sphere in $L_p$ it follows that there exists a sequence of vectors $\{h^m\}_{m=1}^{\infty}$ in the direction of the $x_i$ axis, $h^m \to 0$, such that the sequence $\delta_{h^m} u$ ($m$ sufficiently large) tends weakly in $L_p(G_1)$ to a function $u_i$; and this in every fixed subdomain $G_1$, $\overline{G}_1 \subset G$. Since $\| u_i \|_{L_p(G_1)} \leq C$ for all such subdomains, it follows further that $u_i \in L_p(G)$.

Now, from the definition of weak convergence we find that for all functions $\varphi \in C_0^{\infty}(G)$:

$$\int_G \varphi \, u_i \, dx = \lim_{m \to \infty} \int_G \varphi \cdot \delta_{h^m} u \, dx$$

$$= \lim_{m \to \infty} \int \delta_{-h^m} \varphi \cdot u \, dx = - \int_G D_i \varphi \cdot u \, dx.$$

This shows that $u_i$ is the distribution derivative $D_i u$ in $G$. Since $D_i u \in L_p(G)$ $(i = 1, \ldots, n)$ we conclude from Theorem 3.1 that $u \in H_{1,L_p}(G)$. Clearly, we also have

$$\| D_i u \|_{L_p(G)} \leq C.$$

Next, assume that $j \geq 1$. Let again $\{h^m\}$ be a sequence of vectors in the direction of $x_i$ tending to zero. It is easily seen that $\delta_{h^m} u$ converges to $D_i u$ in $L_p(G_1)$. Assuming without loss of generality that $G_1$ is of class $C^{0,1}$ and applying Lemma 3.1 to the sequence $\{\delta_{h^m} u\}$, it follows that $D_i u \in H_{j,L_p}(G_1)$ and that

$$\| D_i u \|_{j,L_p(G_1)} \leq C \quad (i = 1, \ldots, n).$$

From this and from Lemma 3.2 we conclude that $u \in H_{j+1,L_p}(G_1)$ for any subdomain $G_1$ of class $C^{0,1}$ (and consequently for any subdomain $G_1$, $\overline{G}_1 \subset G$). Since all the distribution derivatives of $u$ of order $\leq j+1$ are functions belonging to $L_p(G)$ it follows from Theorem 3.1 that $u \in H_{j+1,L_p}(G)$. That (3·4) holds is obvious.

By the same argument used to prove Lemma 3.3 for $j = 0$ one obtains

LEMMA 3.3′. *Denote by $\Sigma_R$ the hemisphere $|x| < R$, $x_n > 0$. Let $u$ be a function belonging to $L_p(\Sigma_R)$, $p > 1$. Suppose that there exists a constant $C$ such that for every $R' < R$:*

$$\| \delta^h u \|_{L_p(\Sigma_{R'})} \leq C,$$

*for all sufficiently small vectors h of the form* $h = (h_1, \dots, h_{n-1}, 0)$. *Then the distribution derivatives* $D_i u$ *for* $i = 1, \dots, n-1$ *are functions belonging to* $L_p(\Sigma_R)$ *with* $\| D_i u \|_{L_p(\Sigma_R)} \leq C$.

The following known lemma will be useful.

LEMMA 3.4. *Suppose that G has the cone property. Then, for all functions* $u \in H_{j,L_p}(G)$ $(j \geq 1)$ *and every* $\varepsilon > 0$ *the following inequality holds :*

$$\| u \|_{j-1,L_p(G)} \leq \varepsilon \sum_{|\alpha|=j} \| D^\alpha u \|_{L_p(G)} + C \| u \|_{L_p(G)} \tag{3.5}$$

*where C is a constant depending only on* $\varepsilon, j, p$ *and G.*

Lemma 3.4 for somewhat more regular domains was established by Nirenberg [24] (3). The inequality for domains which have the cone property was proved by Gagliardo [13].

Finally, we conclude this section with the well known integral inequalities of Sobolev [30].

THEOREM 3.2. *Suppose that G has the cone property. Then the functions u belonging to* $H_{j,L_p}(G)$ $(p > 1)$ *satisfy the following relations.*

(i) *If* $p < \frac{n}{j}$ *then* $u \in L_q(G)$ *where q is defined by* $\frac{1}{q} = \frac{1}{p} - \frac{j}{n}$. *Also,*

$$\| u \|_{L_q(G)} \leq \text{Const.} \| u \|_{j,L_p(G)} \tag{3.6}$$

*with a constant depending only on* $n, j, p$ *and G.*

(ii) *If* $p = \frac{n}{j}$ *then* $u \in L_q(G)$ *for every* $1 < q < \infty$ *and* (3.6) *holds.*

(iii) *If* $p > \frac{n}{j}$ *then u is a continuous function (after correction on a null set) such that*

$$\sup_G | u | \leq \text{Const.} \| u \|_{j,L_p(G)}, \tag{3.6$'$}$$

*with the same constant dependence as above.*

REMARK : If the boundary of the domain is somewhat more regular, e. g. if $G$ is of class $C^{0,1}$, one can assert in case (iii) of the theorem that $u$ satisfies a Hölder condition in $\overline{G}$.

## 4. Some lemmas related to elliptic operators with constant coefficients.

Let $A(x, D)$ be a linear differential operator with complex coefficients operating on functions $u(x)$ defined in a domain of $E_n$. Denote by $A'$ the

(3) The analogous one dimensional case is due to Halperin and Pitt.

leading part of $A$, i. e. the part of highest order terms. $A$ is said to be elliptic in the domain if for every point $x$ in the domain the characteristic form $A'(x, \xi) \neq 0$ for all real vectors $\xi = (\xi_1, \ldots, \xi_n) \neq 0$. It is well known that if $n \geq 3$ and $A$ is elliptic then its order is even. This is not necessarily true for $n = 2$.

In this section we shall consider an elliptic operator $A$ of even order $2m$ with constant coefficients and with no lower order terms:

$$(4.1) \qquad A(D) = \sum_{|\alpha|=2m} a_\alpha D^\alpha .$$

A being elliptic there exists a constant $\lambda \geq 1$ such that

$$(4.1)' \qquad \frac{1}{\lambda} | \xi |^{2m} \leq | A(\xi) | \leq \lambda | \xi |^{2m}$$

for all real vectors $\xi$. We term $\lambda$ the ellipticity constant of $A$.

We denote by $x' = (x_1, \ldots, x_{n-1})$ the generic point in $E_{n-1}$ and whenever convenient write $x$ in the form $(x', x_n)$. We also put $D_{x'} = (D_1, \ldots, D_{n-1})$ and $D = (D_{x'}, D_n)$.

Write the operator (4.1) in the form $A(D_{x'}, D_n)$. For a fixed real vector $\xi' = (\xi_1, \ldots, \xi_{n-1}) \neq 0$ consider the roots (in $\xi_n$) of the polynomial $A(\xi', \xi_n)$. If $n \geq 3$ the ellipticity of $A$ implies the exactly half the roots possess a positive imaginary part (see [3]). This is not necessarily true for $n = 2$ if the coefficients are not real. In general we shall say that $A$ satisfies the « roots condition » if for every fixed real vector $\xi' \neq 0$ the polynomial $A(\xi', \xi_n)$ has exactly $m$ roots with a positive imaginary part.

The following two lemmas are basic for the proof of regularity in $L_p$ of weak solutions of elliptic equations. The first rather known lemma will be used to establish interior regularity (and $L_p$ estimates) of weak solutions of elliptic equations and overdetermined elliptic systems. The second lemma will be used to establish regularity at the boundary of weak solutions of the Dirichlet problem. In both lemmas $A$ will stand for the elliptic operator (4.1) and $p$ will denote a number $> 1$. In Lemma 4.2 we shall assume in addition, if $n = 2$, that $A$ satisfies the « roots condition » introduced above. We shall denote by $S_R$ the sphere $|x| < R$ and by $\Sigma_R$ the half sphere $|x| < R$, $x_n > 0$.

LEMMA 4.1. *Given a function* $f \in C_0^\infty(S_R)$ *there exists a function* $v \in C^\infty(\bar{S}_R)$ *such that*

$$(4.2) \qquad Av = f \quad in \quad \bar{S}_R$$

*and*

$$\| v \|_{2m, L_p(S_R)} \leq C \| f \|_{L_p(S_R)}, \tag{4.2)'}$$

*where $C$ is some constant depending only on $n, m, p, R$ and $\lambda$ (but not on $f$ or $v$).*

LEMMA 4.2. *Given a function $f \in C_0^\infty(\Sigma_R)$ there exists a function $v \in C^\infty(\bar{\Sigma}_R)$ such that*

$$\begin{cases} Av = f & \text{in } \bar{\Sigma}_R \\ D_n^j v = 0 & \text{for } x_n = 0 \quad (|x| \leq R), \quad j = 0, \ldots, m-1, \end{cases} \tag{4.3}$$

*and*

$$\| v \|_{2m, L_p(\Sigma_R)} \leq C \| f \|_{L_p(\Sigma_R)}, \tag{4.3)'}$$

*where $C$ is some constant depindig only on $n, m, p, R$ and $\lambda$.*

To establish Lemma 4.1 we simply define

$$v(x) = \int_{S_R} f(y) F(x - y) \, dy \tag{4.4}$$

where $F(x-y)$ is a suitable chosen fundamental solution of $A$ with pole at $x = y$. It is well known (e. g. F. John [16]) that there exists a fundamental solution having the form:

$$F(x) = |x|^{2m-n} \psi\left(\frac{x}{|x|}\right) + P(x) \log |x|, \tag{4.5}$$

where $P(x)$ is a polynomial of degree $2m - n$ if $n$ is even, $2m \geq n$, and $P(x)$ is zero otherwise; $\psi(y)$ is an analytic function defined on $|y| = 1$. From (4.5) it follows that

$$|D^\alpha F(x)| \leq \text{Const.} \, |x|^{2m-n-|\alpha|} \tag{4.6}$$

for (i) $|\alpha| \geq 0$, in case $n$ is odd or $n$ is even and greater than $2m$; (ii) $|\alpha| > 2m - n$ if $n$ is even and not greater than $2m$. If $n$ is even and $0 \leq |\alpha| \leq 2m - n$, then

$$|D^\alpha F(x)| \leq \text{Const.} \, |x|^{2m-n-|\alpha|} (1 + |\log |x||). \tag{4.6)'}$$

Inspection of the explicit formulas for the fundamental solution (in [16]) shows that the constants in (4.6) and (4.6)' depend only on $m, n, \lambda$ and

$|\alpha|$. Furthermore, it is easily established that $D^\alpha F(x)$ for $|\alpha| = 2m$ is a homogeneous function of degree $-n$ with a zero mean on the sphere $|x| = 1$.

Choosing a proper normalization of $F$, the function $v$ defined by (4.4) is infinitely differentiable and satisfies (4.2). Furthermore, from the properties of the fundamental solution mentioned above and from the well known theorem of Calderon and Zygmund [8] on convolution transforms with singular kernels, it follows readily that

$$\| v \|_{2m, L_p(S_R)} \leq C \| f \|_{L_p(S_R)}$$

where $C$ is a constant dependig only $n, m, p, R$ and $\lambda$. Hence, the function $v$ defined by (4.4) answers all the requirements of Lemma 4.1.

The proof Lemma 4.2 is more involved and depends on the solution of the Dirichlet problem for $A$ in *a* half space and related $L_p$ estimates.

We shall denote by $E_n^+$ the half space $x_n > 0$. In its simplest form the Dirichlet problem for $A$ in $E_n^+$ is the following

PROBLEM: *Given functions $\varphi_1(x'), \ldots, \varphi_m(x')$, infinitely differentiable and of compact support in $E_{n-1}$, find an infinitely differentiable function $u(x', x_n)$ in $\overline{E}_n^+$ such that*

$$\tag{4.7} \begin{cases} Au = 0 \quad in \quad \overline{E}_n^+, \\ D_n^{j-1} u(x', 0) = \varphi_j(x') \quad for \quad j = 1, \ldots, m. \end{cases}$$

This problem (a special case among a whole class of boundary value problems) was solved in [3] ([4]), where it was shown that there exist kernels $K_j(x', x_n)$ $(j = 1, \ldots, m)$, defined and infinitely differentiable in $\overline{E}_n^+$ except for the origin, such that a solution of (4.7) is given by the formula:

$$\tag{4.8} u(x', x_n) = \sum_{j=1}^{m} \int_{E_{n-1}} \varphi_j(y') K_j(x' - y', x_n)\, dy'.$$

We mention the following properties of the kernels $K_j$ also established [3]. Let $q$ be a non-negative integer having the same parity as $n-1$. The kernel $K_j$ admits a representation of the form

$$\tag{4.9} K_j(x', x_n) = \Delta_{x'}^{\frac{1}{2}(n-1+q)} K_{j,q}(x', x_n)$$

---

([4]) For $n = 2$ and $A$ real the solution was given in [1].

where $\Delta_{x'}$ is the Laplacean $\sum_{i=1}^{n-1} D_i^2$, and $K_{j,q}$ are certain kernels which are infinitely differentiable in $\overline{E}_n^+$ except for the origin which, moreover, satisfy the following inequalities in $E_n^+$:

$$(4.10) \qquad | D^\alpha K_{j,q}(x) | \leq \text{Const.} \, | x |^{j-1+q-|\alpha|} (1 + | \log | x || )$$

for $| \alpha | \leq j - 1 + q$, and

$$(4.10)' \qquad | D^\alpha K_{j,q}(x) \leq \text{Const.} \, | x |^{j-1+q-|\alpha|}$$

for $| \alpha | \geq j + q$, where the constants in (4.10) and (4.10)′ depend only on $n, m, q, | \alpha |$ and the ellipticity constant $\lambda$.

Let, now, $w(x', x_n)$ be an infinitely differentiable function with compact support in $\overline{E}_n^+$. By the preceding a solution $u \in C^\infty(\overline{E}_n^+)$ of the Dirichlet problem

$$(4.11) \qquad Au = 0 \quad \text{in} \quad \overline{E}_n^+,$$

$$D_n^{j-1} u(x', 0) = D_n^{j-1} w(x', 0) \quad \text{for} \quad j = 1, \ldots, m,$$

is given by the formula

$$(4.11)' \qquad u(x', x_n) = \sum_{j=1}^{m} \int_{E_{n-1}} D_n^{j-1} w(y', 0) \cdot K_j(x' - y', x_n) \, dy' .$$

Moreover, we have

LEMMA 4.3. *The solution $u$ satisfies the following inequality in $L_p, p > 1$:*

$$(4.12) \qquad \sum_{|\alpha|=2m} \| D^\alpha u \|_{L_p(E_n^+)} \leq c_0 \sum_{|\alpha|=2m} \| D^\alpha w \|_{L_p(E_n^+)} ,$$

*where $c_0$ is a constant depending only on $m, n, p$ and $\lambda$. If, in addition, the support of $w$ is contained in the halfsphere $\Sigma_R$ then*

$$(4.12)' \qquad \| u \|_{2m, L_p(\Sigma_R)} \leq C_0 \| w \|_{2m, L_p(\Sigma_R)}$$

*where $C_0$ is a constant depending only on $n, m, p, \lambda$ and $R$.*

Lemma 4.3. was proved (essentially) in [3] (compare also Koselev [17] for the $L_p$ estimates involved). For the sake of completeness we shall present a somewhat simplified version of the proof later on. It is with the aid of this lemma that we shall now give the

*Proof of Lemma* 4.2. Extend the function $f(x)\,(f \in C_0^\infty(\Sigma_R))$ as zero outside $\Sigma_R$. Denote by $\zeta_R$ some fixed infinitely differentiable function such that $\zeta_R \equiv 1$ for $|x| \leq R$, $\zeta_R \equiv 0$ for $|x| \geq 2R$. Define:

$$w(x) = \zeta_R(x) \int_{E_n} f(y)\, F(x-y)\, dy\,, \tag{4.13}$$

where $F$ is the fundamental solution of $A$ introduced before. Clearly, $w$ is infinitely differentiable, $w \equiv 0$ for $|x| \geq 2R$, and

$$\begin{cases} Aw = f \quad \text{for} \quad |x| \leq R\,, \\ \| w \|_{2m, L_p(\Sigma_{2R})} \leq C_1 \| f \|_{L_p(\Sigma_R)}\,, \end{cases} \tag{4.14}$$

where $C_1$ is a constant depending only on $n, m, p, \lambda$ and $R$. Let, now, $u$ be the solution of the Dirichlet problem (4.11) given by (4.11)′ with $w$ defined by (4.13). Put:

$$v = w - u\,.$$

Then, $v$ has all the properties required by Lemma 4.2. Indeed, $v \in C^\infty(\overline{E}_n^+)$. By (4.14) and (4.11):

$$\begin{cases} Av = Aw = f \quad \text{for} \quad x \in \overline{\Sigma}_R\,, \\ D_n^{j-1} v = D_n^{j-1} w - D_n^{j-1} u = 0 \quad \text{for} \quad x_n = 0\,, j = 1\,, \dots, m\,. \end{cases}$$

Finally, using the estimate (4.12)′ of Lemma 4.3 and (4.14), we get

$$\| v \|_{2m, L_p(\Sigma_R)} \leq \| w \|_{2m, L_p(\Sigma_R)} + \| u \|_{2m, L_p(\Sigma_R)}$$

$$\leq \| w \|_{2m, L_p(\Sigma_R)} + C_0 \| w \|_{2m, L_p(\Sigma_{2R})} \leq C \| f \|_{L_p(\Sigma_R)}\,.$$

where $C_0$, $C$ are constants depending only on $n, m, p, \lambda$ and $R$. This establishes the lemma.

We shall conclude the section with a proof of Lemma 4.3 based on the properties of the kernels $K_j$ mentioned before. We shall need first the following

Sublemma: *Let $G(x) = G(x', x_n)$ be a kernel, defined and measurable in the half space $E_n^+$ such that*

$$| G(x) | \leq C | x |^{-n}\,, \tag{4.15}$$

*for some constant $C$. For $v \in L_p(E_n^+)$, $p > 1$, consider the transform*

$$u(x', x_n) = \iint_{E_n^+} v(y', y_n)\, G(x' - y', x_n + y_n)\, dy'\, dy_n . \tag{4.16}$$

*Then, $u \in L_p(E_n^+)$ and*

$$\| u \|_{L_p(E_n^+)} \le \gamma\, C \| v \|_{L_p(E_n^+)} ; \tag{4.16'}$$

*where $\gamma$ is a constant depending only on $n$ and $p$.*

*Proof:* Set

$$\begin{cases} M(x) = \ \ | x |^{-n} & \text{for} \quad x_n > 0 , \\ M(x) = -| x |^{-n} & \text{for} \quad x_n < 0 . \end{cases}$$

Extend $v$ as zero for $x_n \le 0$. Then, for $x_n > 0$:

$$| u(x) | \le C \iint_{E_n^+} | v(y', y_n) |\, M(x' - y', x_n + y_n)\, dy'\, dy_n \tag{4.17}$$

$$= C \iint_{E_n} | v(y', -y_n) |\, M(x - y)\, dy'\, dy_n .$$

Now, $M(x)$ is an odd homogeneous kernel of degree $-n$ bounded on $|x| = 1$. Hence, we are in a position to apply the Calderon-Zygmund theorem [8] to the last integral (4.17), from which it follows readily that

$$\| u \|_{L_p(E_n^+)} \le \gamma\, C \| v \|_{L_p(E_n)} = \gamma\, C \| v \|_{L_p(E_n^+)} ,$$

$\gamma$ depending only on $n$ and $p$. This proves the sublemma.

To prove Lemma 4.3 we shall first transform formula (4.11)′. To this end note that (integrating by parts with respect to $y_n$)

$$\int_{E_{n-1}} D_n^{j-1} w(y', 0) \cdot K_j(x' - y', x_n)\, dy' \tag{4.18}$$

$$= - \iint_{E_n^+} D_n^j w(y', y_n) \cdot K_j(x' - y', x_n + y_n)\, dy'\, dy_n -$$

$$- \iint_{E_n^+} D_n^{j-1} w(y', y_n) \cdot D_n K_j(x' - y', x_n + y_n)\, dy'\, dy_n ,$$

where here and in the following all differential operators under the integral sign act on the $y$ variable unless otherwise indicated by a subscript. Summing (4.18) over $j = 1, \dots, m$ we obtain for the solution $u$ of (4.11) the representation:

$$u(x', x_n) = \sum_{j=0}^{m} \iint_{E_n^+} D_n^j w(y', y_n) \cdot \widetilde{K}_j(x' - y', x_n + y_n)\, dy'\, dy_n , \tag{4.19}$$

where

$$\begin{cases} \widetilde{K}_j = -K_j - D_n K_{j+1} , & j = 1, \dots, m-1 , \\ \widetilde{K}_0 = -D_n K_1 , \quad \widetilde{K}_m = K_m . \end{cases} \tag{4.20}$$

Using (4.19) and (4.9) we observe that if $q$ is a non-negative integer having the same parity as $n - 1$ then

$$\widetilde{K}_j(x', x_n) = \Delta_{x'}^{\frac{1}{2}(n-1+q)} \widetilde{K}_{j,q}(x', x_n), \tag{4.21}$$

where $\widetilde{K}_{j,q}$ are kernels given by

$$\begin{cases} \widetilde{K}_{j,q} = -K_{j,q} - D_n K_{j+1,q} , & j = 1, \dots, m-1 , \\ \widetilde{K}_{0,q} = -D_n K_{1,q} , \quad \widetilde{K}_{m,q} = K_{m,q} . \end{cases} \tag{4.20$'$}$$

From (4.20)′ it is readily seen that the inequalities (4.10)-(4.10)′, satisfied by the kernels $K_{j,q}$, are also satisfied by the kernels $\widetilde{K}_{j,q}$.

Put:

$$u_j(x) = \iint_{E_n^+} D_n^j w(y) \cdot \widetilde{K}_j(x' - y', x_n + y_n)\, dy \quad (x_n > 0) , \tag{4.22}$$

so that by (4.19) $u = \sum_0^m u_j$. To establish the lemma it will suffice to show that the inequalities (4.12) and (4.12)′ hold for $u_j$. We shall prove this for $j$ odd. The proof for $j$ even is similar.

Choose $q = 2m + n + 1$. From (4.22) and (4.21) we obtain after obvious integration by parts with respect to $y'$:

$$u_j(x) = \iint\limits_{E_n^+} D_n^j w(y) \cdot \Delta_{y'}^{\frac{1}{2}(n-1+q)} \widetilde{K}_{j,q}(x'-y', x_n+y_n)\, dy \tag{4.23}$$

$$= \iint\limits_{E_n^+} D_n^j \Delta_{y'}^{m-\frac{1}{2}(j+1)} w(y) \cdot \Delta_{y'}^{n+\frac{1}{2}(j+1)} \widetilde{K}_{j,q}(x'-y', x_n+y_n)\, dy$$

$$= -\sum_{i=1}^{n-1} \iint\limits_{E_n^+} D_i D_n^j \Delta_{y'}^{m-\frac{1}{2}(j+1)} w \cdot D_i \Delta_{y'}^{n+\frac{1}{2}(j-1)} \widetilde{K}_{j,q}(x'-y', x_n+y_n)\, dy .$$

Differentiating (4.23) we thus obtain:

$$-D^\alpha u_j(x) = \tag{4.23'}$$

$$= \sum_{i=1}^{n-1} \iint\limits_{E_n^+} D_i D_n^j \Delta_{y'}^{m-\frac{1}{2}(j+1)} w \cdot D_i \Delta_{y'}^{n+\frac{1}{2}(j-1)} D_x^\alpha \widetilde{K}_{j,q}(x'-y', x_n+y_n)\, dy .$$

Suppose, first, that $|\alpha| = 2m$. Using the estimates (4.10)-(4.10)′ which, as was pointed out before, are also satisfied by the kernels $\widetilde{K}_{j,q}$ $(q=2m+n+1)$, we find that

$$| D_i \Delta_{x'}^{n+\frac{1}{2}(j-1)} D^\alpha \widetilde{K}_{j,q}(x', x_n) | \leq c\, |x|^{-n} \tag{4.24}$$

with a constant $c$ depending only on $n$, $m$ and $\lambda$. Thus, applying the Sublemma to a typical integral of (4.23)′ it follows readily that

$$\| D^\alpha u_j \|_{L_p(E_n^+)} \leq \gamma_1\, c \sum_{|\beta|=2m} \| D^\beta w \|_{L_p(E_n^+)}$$

where $\gamma_1$ depends only on $n$ and $p$. This yields (4.12).

Suppose, now, that $0 \leq |\alpha| \leq 2m-1$. From (4.10)-(4.10)′ one finds readily that in this case

$$| D_i \Delta_{x'}^{n+\frac{1}{2}(j-1)} D^\alpha \widetilde{K}_{j,q}(x', x_n) | \leq \text{Const.} ( |x|^{-(n-1)} + |x|^{2m}) \tag{4.25}$$

with a constsnt depending only on $n$, $m$ and $\lambda$. If, furthermore, the support of $w$ is contained in $\overline{\Sigma}_R$ it follows easily from (4.23) and (4.25) that for $|\alpha| < 2m$:

$$\| D^\alpha u_j \|_{L_p(\Sigma_R)} \leq C_1 \sum_{|\beta|=2m} \| D^\beta w \|_{L_p(\Sigma_R)} \tag{4.26}$$

where $C_1$ is a constant depending only on $n$, $m$, $\lambda$, $p$ and $R$. Since by the preceding (4.26) (with a suitable constant) holds also for $|\alpha| = 2m$ we conclude that the functions $u_j$ (and consequently $u$) satisfy (4.12)'. This completes the proof of Lemma 4.3.

## 5. Preliminary regularity lemmas.

In this section we begin with the discussion of the regularity problems of weak solutions of elliptic equations in the framework of the $L_p$ theory. We shall discuss both the problem of interior regularity (also for weak solutions of overdetermined elliptic systems), and the problem of regularity at the boundary for weak solutions of the Dirichlet problem,

We consider a linear elliptic differential operator $A$ of order $2m$ (variable complex coefficients) defined in $\overline{G}$:

$$A(x, D) = \sum_{|\alpha| \leq 2m} a_\alpha(x) D^\alpha . \tag{5.1}$$

We denote by $A'$ the leading part of $A$ and by $\lambda$ some constant $\geq 1$ (el lipticity constant) such that

$$\frac{1}{\lambda} |\xi|^{2m} \leq |A'(x, \xi)| \leq \lambda |\xi|^{2m} \tag{5.1$'$}$$

for all real vectors $\xi$ and $x \in \overline{G}$. We introduce the following

DEFINITION 5.1. *The coefficients of $A$ will be said to satisfy* Condition $\{j; K\}$ *(in $\overline{G}$), $j$ being a positive integer and $K > 0$, if*

(i) $$a_\alpha \in C^{|\alpha|+j-2m}(\overline{G}) \quad \text{for} \quad |\alpha| > 2m - j ,$$

*whereas the remaining coefficients are measurable bounded functions in $G$.*
(ii) *The following inequalities hold in $G$:*

$$|D^\beta a_\alpha| \leq K \quad \textit{for} \quad |\alpha| > 2m - j , \quad |\beta| \leq |\alpha| + j - 2m ,$$

*and*

$$|a_\alpha| \leq K \quad \textit{for} \quad |\alpha| \leq 2m - j .$$

We also introduce the scalar product notation $(f, g)_G$ to be used from now on throughout the paper:

$$(f, g)_G = \int_G f \bar{g} \, dx . \tag{5.2}$$

Here $f$ and $g$ are two functions defined in $G$ such that the integral (5.2) makes sense.

In this and in the following section the domains of definition $G$ will be either the sphere $S_R$ ($|x| < R$) or the hemisphere $\Sigma_R$ ($|x| < R, x_n > 0$), We shall denote by $\partial_1 \Sigma_R$ the part of the boundary $\partial \Sigma_R$ situated on the hyperplane $x_n = 0$, and by $\partial_2 \Sigma_R$ the part of $\partial \Sigma_R$ situated on $|x| = R$ ($\partial \Sigma_R = \partial_1 \Sigma_R \cup \partial_2 \Sigma_R$). We also recall that by $\delta_h$ we denote the difference quotient operator:

$$\delta_h u = \frac{u(x+h) - u(x)}{|h|},$$

$h = (h_1, \ldots, h_n)$ being a real vector $\neq 0$.

We shall now state two basic regularity lemmas.

LEMMA 5.1. *Let $A$ be an elliptic differential operator of order $2m$ defined in $\bar{S}_R$, with coefficients satisfying* Condition $\{1 ; K\}$. *Let, further, $u$ be a function belonging to $L_p(S_R)$, $p > 1$, such that for all functions $\varphi \in C_0^\infty(S_R)$ the following inequality holds:*

$$|(u, A\varphi)_{S_R}| \leq C \| \varphi \|_{2m-1, L_{p'}(S_R)}, \tag{5.3}$$

*where $p'$ is the exponent conjugate to $p$: $\dfrac{1}{p} + \dfrac{1}{p'} = 1$, and $C$ is a constant. Then, there exists a positive number $r_0 < R$ and a constant $c_0$* such that

$$\| \delta_h u \|_{L_p(S_{r_0})} \leq c_0 (C + \| u \|_{L_p(S_R)}), \tag{5.4}$$

*for all sufficiently small vectors $h$. Both $r_0$* and $c_0$ *depend only on $n, m, p, R, K$ and the ellipticity constant $\lambda$.*

LEMMA 5.2. *Let $A(x, D)$ be an elliptic differential operator of order $2m$ defined in $\bar{\Sigma}_R$, with coefficients satisfying* Condition $\{1 ; K\}$. *If $n = 2$ assume also that $A'(0, D)$ satisfies the «roots condition» introduced in § 4. Let, further, $u$ be a function belonging to $L_p(\Sigma_R)$, $p > 1$, such that*

$$|(u, A\varphi)_{\Sigma_R}| \leq C \| \varphi \|_{2m-1, L_{p'}(\Sigma_R)} \tag{5.5}$$

*for all functions* $\varphi \in C^\infty(\overline{\Sigma}_R)$ *satisfying the boundary conditions :*

$$(5.6)\qquad \begin{cases} D_n^j \varphi = 0 & \text{on } \partial_1 \Sigma_R, \quad j = 0, \dots, m-1. \\ \varphi \equiv 0 & \text{in a neighborhood of } \partial_2 \Sigma_R. \end{cases}$$

*Then, there exist a positive nomber* $r_0 < R$ *and a constant* $c_0$, *having the same dependence as in Lemma* 5.1, *such that*

$$(5.7)\qquad \| \delta_h u \|_{L_p(\Sigma_{r_0})} \leq c_0 (C + \| u \|_{L_p(\Sigma_R)}),$$

*for all sufficiently small vectors h parallel to the hyperplane* $x_n = 0$.

*Proof of Lemma* 5.1 *and Lemma* 5.2 : We shall prove both lemmas at the same time. In the sequel $\sigma_r$ will denote the sphere $S_r$ in the case of Lemma 5.1, and the hemisphre $\Sigma_r$ in the case of Lemma 5.2.

By our assumption the coefficients of $A$ are measurable functions bounded by $K$ in $\sigma_R$. Moreover, the highest order coefficients possess first order derivatives also bounded by $K$. Without loss of generality we may assume in the following that $A$ contains no lower order terms. Indeed in the general case let $A = A' + A''$ where $A'$ is the leading part and $A''$ contains only terms of order $\leq 2m - 1$. By Hölder's inequality

$$|(u, A'' \varphi)_{\sigma_R}| \leq K \| u \|_{L_p(\sigma_R)} \cdot \| \varphi \|_{2m-1, L_{p'}(\sigma_R)}.$$

Consequently the operator $A$ could be replaced by $A'$ which satisfies the conditions of the lemmas with $C$ replaced by

$$C' = C + K \| u \|_{L_p(\sigma_R)}.$$

Let $r$ be a positive number $\leq R/6$ to be fixed later on, and let $\zeta_r$ be a real $C^\infty$ function such that $\zeta_r \equiv 1$ for $|x| \leq r/3$, $\zeta \equiv 0$ for $|x| \geq \frac{2}{3} r$. Set :

$$(5.8)\qquad u_r = \zeta_r u, \qquad x \in \overline{\sigma}_R,$$

and extend $u_r$ as zero outside $\overline{\sigma}_R$. Let, further, $v$ be and arbitrary function belonging to $C^\infty(\overline{\sigma}_r)$ which in the case of Lemma 5.2 ($\sigma_r = \Sigma_r$) also satisfies the boundary conditions

$$(5.9)\qquad D_n^j v = 0 \quad \text{for} \quad x_n = 0, \qquad j = 0, \dots, m-1.$$

We have :

$$(5.10)\quad (u, A(\zeta_r v))_{\sigma_r} = (u, \zeta_r A v)_{\sigma_r} + (u, B v)_{\sigma_r} = (u_r, A v)_{\sigma_r} + (u, B v)_{\sigma_r},$$

where $B$ is a linear differential operator of order $2m - 1$ with coefficients which are linear combinations in $a_\alpha D^\beta \zeta_r$. Hence, making use of Hölder's inequality, we obtain

$$(5.10)' \qquad |(u, A(\zeta_r v))_{\sigma_r} - (u_r, A v)_{\sigma_r}| \leq c_2 \| u \|_{L_p(\sigma_r)} \| v \|_{2m-1, L_{p'}(\sigma_r)},$$

where here and in the following $c_2, c_3, \ldots$, denote constants depending only on $n, m, p, R, K, \lambda$, and $r$.

Consider now the function $\varphi = \zeta_r v$ extended as zero outside $\bar{\sigma}_r$. $\varphi \in C_0^\infty(S_R)$ in the case of Lemma 5.1, while $\varphi \in C^\infty(\bar{\Sigma}_R)$ and satisfies the boundary conditions (5.6) in the case of Lemma 5.2. Hence, applying the inequality (5.3) in the first case, and the inequality (5.5) in the second, we conclude that

$$(5.11) \qquad |(u, A(\zeta_r v))_{\sigma_r}| \leq c_3 C \| v \|_{2m-1, L_{p'}(\sigma_r)}.$$

Combining (5.11) with (5.10)′ we thus get:

$$(5.12) \qquad |(u_r, A v)_{\sigma_r}| \leq c_4 \| v \|_{2m-1, L_{p'}(\sigma_r)} (C + \| u \|_{L_p(\sigma_r)})$$

for all functions $v \in C^\infty(\bar{\sigma}_r)$ which in the case of Lemma 5.2 also satisfy the boundary conditions (5.9).

Next, let $h$ be an arbitrary vector such that $0 < |h| < r/6$. In the case of Lemma 5.2 $h$ is restricted, in addition, to be of the form $h = (h_1, \ldots, h_{n-1}, 0)$. Define the function

$$(5.13) \qquad f_h(x) = |\delta_h u_r|^{p-1} \operatorname{sign}(\delta_h u_r) \text{ (5)}.$$

Then, $f_h \in L_{p'}(\sigma_R)$,

$$(5.13)' \qquad \| f_h \|_{L_{p'}(\sigma_R)} = \| \delta_h u_r \|^{p-1}_{L_p(\sigma_R)},$$

and the support of $f_h$ is contained in $\bar{\sigma}_{R/3}$. Let, further, $\tilde{f}_h$ be a $C^\infty$ function with support in $\sigma_{R/3}$ such that

$$(5.14) \qquad \| \tilde{f}_h - f_h \|_{L_{p'}(\sigma_R)} \leq \frac{1}{3} \| f_h \|_{L_{p'}(\sigma_R)}.$$

---

(5) As usual: sign $z = \frac{z}{|z|}$ if $z \neq 0$, sign $0 = 0$.

We now make use of the lammas established in § 4. Thus, in the case of Lemma 5.1 apply Lemma 4.1 to the elliptic operator $A^0 = A(0, D)$, function $f = \tilde{f}_h$ and exponent $p'$. There exists by the lemma a function $v_h \in C^\infty(\bar{S}_R)$ such that

$$A^0 v_h = \tilde{f}_h \quad \text{in} \quad \bar{S}_R, \tag{5.15}$$

and

$$\| v_h \|_{2m, L_{p'}(S_R)} \leq \gamma \, | \tilde{f}_h |_{L_{p'}(S_R)}, \tag{5.15$'$}$$

where $\gamma$ is a constant depending only on $n$, $m$, $p$, $R$ and $\lambda$.

Similarly, in the case of Lemma 5.2, applying Lemma 4.2 it follows that there exists a function $v_h \in C^\infty(\bar{\Sigma}_R)$ such that

$$\begin{cases} A^0 v_h = \tilde{f}_h & \text{in} \quad \Sigma_R \\ D_n^j v_h = 0 & \text{on} \quad \partial_1 \Sigma_R \; j = 0, \ldots, m-1, \end{cases} \tag{5.16}$$

and

$$\| v_h \|_{2m, L_{p'}(\Sigma_R)} \leq \gamma \, \| \tilde{f}_h \|_{L_{p'}(\Sigma_R)}, \tag{5.16$'$}$$

where $\gamma$ is a constant having the same dependence as above.

Using (5.13)′ and (5.14) we observe that in both cases

$$\| v_h \|_{2m, L_{p'}(\sigma_R)} \leq 2\gamma \, \| \delta_h u_r \|^{p-1}_{L_p(\sigma_R)}. \tag{5.17}$$

Consider the function $\delta_{-h} v_h$. It is a well defined infinitely differentiable function in $\bar{\sigma}_r$. Moreover, in the case of Lemma 5.2 it also satisfies the boundary conditions (5.9). Hence, applying (5.12) to $v = \delta_{-h} v_h$ we have:

$$| (u_r, A(\delta_{-h} v_h))_{\sigma_r} | \leq c_4 \| \delta_{-h} v_h \|_{2m-1, L_{p'}(\sigma_r)} (C + \| u \|_{L_p(\sigma_r)}). \tag{5.18}$$

Also, one checks easily that

$$\| \delta_{-h} v_h \|_{2m-1, L_{p'}(\sigma_r)} \leq N \| v_h \|_{2m, L_{p'}(\sigma_R)}, \tag{5.19}$$

where $N$ is a constant depending only on $n$ (one can actually take $N = n$). Combining (5.18), (5.19) and (5.17) we thus get

$$| (u_r, A(\delta_{-h} v_h))_{\sigma_r} | \leq c_5 \| \delta_h u_r \|^{p-1}_{L_p(\sigma_r)} (C + \| u \|_{L_p(\sigma_r)}). \tag{5.20}$$

Put:

$$A_h = \frac{A(x+h, D) - A(x, D)}{|h|} = \sum_{|\alpha|=2m} \frac{a_\alpha(x+h) - a_\alpha(x)}{|h|} D^\alpha.$$

Using the relation $f\,\delta_{-h}\,g = \delta_{-h}(fg) - \delta_{-h} f \cdot g(x-h)$, and noting that the support of $u_r$ is contained in $\overline{\sigma}_{2r/3}$ while $|h| < r/6$, we readily obtain:

(5.21) $$(u_r, A(\delta_{-h} v_h))_{\sigma_r} = (u_r, \delta_{-h}(A v_h))_{\sigma_r} - (u_r, A_{-h} v_h(x-h))_{\sigma_r} =$$

$$= (\delta_h u_r, A v_h)_{\sigma_r} - (u_r, A_{-h} v(x-h))_{\sigma_r}.$$

Since the coefficients of $A_h$ are bounded by $n\,K$ we have (using 5.17):

(5.22) $$|(u_r, A_{-h} v_h(x-h))_{\sigma_r}| \leq c_6 \| u_r \|_{L_p(\sigma_r)} \| v_h \|_{2m, L_{p'}(\sigma_R)} \leq$$

$$\leq c_7 \| u_r \|_{L_p(\sigma_r)} \| \delta_h u_r \|^{p-1}_{L_p(\sigma_r)}.$$

Thus, combining (5.20), (5.21) and (5.22), we get

(5.23) $$|(\delta_h u_r, A v_h)_{\sigma_r}| \leq |(u_r, A(\delta_{-h} v_h))_{\sigma_r}| + |(u_r, A_{-h} v_h(x-h))_{\sigma_r}| \leq$$

$$\leq c_8 \| \delta_h u_r \|^{p-1}_{L_p(\sigma_r)} (C + \| u \|_{L_p(\sigma_r)}).$$

Write

(5.24) $$(\delta_h u_r, A v_h)_{\sigma_r} = (\delta_h u_r, A^0 v_h)_{\sigma_r} + (\delta_h u_r, (A - A^0) v_h)_{\sigma_r}.$$

Using (5.15) (resp. (5.16)) we have:

(5.25) $$(\delta_h u_r, A^0 v_h)_{\sigma_r} = (\delta_h u_r, \tilde{f}_h)_{\sigma_r} = (\delta_h u_r, f_h)_{\sigma_r} + (\delta_h u_r, \tilde{f}_h - f_h)_{\sigma_r}.$$

By (5.13):

(5.25)′ $$(\delta_h u_r, f_h)_{\sigma_r} = \| \delta_h u_r \|^p_{L_p(\sigma_r)}.$$

Also, from Hölder's inequality and (5.14) we get

(5.25)″ $$|(\delta_h u_r, \tilde{f}_h - f_h)_{\sigma_r}| \leq \| \delta_h u_r \|_{L_p(\sigma_r)} \| \tilde{f}_h - f_h \|_{L_{p'}(\sigma_r)} \leq$$

$$\leq \frac{1}{3} \| \delta_h u_r \|^p_{L_p(\sigma_r)}.$$

Thus, combining (5.25), (5.25)′ and (5.25)″ we obtain

$$(5.26)\qquad |(\delta_h u_r, A^0 v_h)_{\sigma_r}| \geq$$

$$\geq (\delta_h u_r, f_h)_{\sigma_r} - |(\delta_h u_r, \tilde{f}_h - f_h)_{\sigma_r}| \geq \frac{2}{3} \| \delta_h u_r \|^p_{L_p(\sigma_r)} .$$

Now, it follows from our assumption that the coefficients of $A - A^0$ are bounded by $n K r$ in $\sigma_r$. From this and from (5.17) we get, using once more Hölder's inequality:

$$(5.27)\qquad |(\delta_h u_r, (A - A^0) v_h)_{\sigma_r}| \leq \| \delta_h u_r \|_{L_p(\sigma_r)} \| (A - A^0) v_h \|_{L_{p'}(\sigma_r)} \leq$$

$$\leq r K n^{2m+1} \| \delta_h u_r \|_{L_p(\sigma_r)} \| v_h \|_{2m, L_{p'}(\sigma_r)} \leq 2 \gamma r K n^{2m+1} \| \delta_h u_r \|^p_{L_p(\sigma_r)} .$$

We shall fix now $r$ choosing

$$(5.28)\qquad r = \min \left( \frac{R}{6}, \frac{1}{6 \gamma K n^{2m+1}} \right).$$

With this choice of $r$ we obtain from (5.27) and (5.26):

$$(5.29)\qquad |(\delta_h u_r, A v_h)_{\sigma_r}| \geq |(\delta_h u_r, A^0 v_h)_{\sigma_r}| -$$

$$- |(\delta_h u_r, (A - A^0) v_h)_{\sigma_r}| \geq \frac{1}{3} \| \delta_h u_r \|^p_{L_p(\sigma_r)} .$$

Finally, from (5.23) and (5.29) we get

$$\frac{1}{3} \| \delta_h u_r \|^p_{L_p(\sigma_r)} \leq c_8 \| \delta_h u_r \|^{p-1}_{L_p(\sigma_r)} (C + \| u \|_{L_p(\sigma_r)}),$$

or,

$$(5.30)\qquad \| \delta_h u_r \|_{L_p(\sigma_r)} \leq 3 c_8 (C + \| u \|_{L_p(\sigma_r)}) .$$

If we now choose $r_0 = \frac{r}{6}$, $c_0 = 3 c_8$, and note that $\delta_h u_r = \delta_h u$ for $|x| \leq r_0$, $|h| \leq r_0$, we obtain from (5.30):

$$\| \delta_h u \|_{L_p(\sigma_{r_0})} \leq \| \delta_h u_r \|_{L_p(\sigma_r)} \leq c_0 (C + \| u \|_{L_p(\sigma_R)}) ,$$

for all $h$ sufficiently small ($h$ parallel to $x_n = 0$ in the case of Lemma 5.2). This establishes the lemmas.

Lemma 5.1 and Lemma 5.2 yield (respectively) the following corollaries.

COROLLARY 5.1. *Suppose that the conditions of Lemma 5.1 hold. Then,* $u \in H^{\text{loc.}}_{1,L_p}(S_R)$. *Moreover, for every $R' < R$ the following inequality holds:*

$$\| u \|_{1,L_p(S_{R'})} \leq c_1 (C + \| u \|_{L_p(S_R)}) \tag{5.31}$$

*where $c_1$ is a constant depending only on $m, n, p, \lambda, K, R$ and $R'$.*

COROLLARY 5.2. *Suppose that the conditions of Lemma 5.2. hold. Then, for every $R' < R$ the distribution derivatives $D_i u$, for $i = 1, \dots, n-1$, are functions belonging to $L_p(\Sigma_{R'})$ such that*

$$\sum_{i=1}^{n-1} \| D_i u \|_{L_p(\Sigma_{R'})} \leq c_1 (C + \| u \|_{L_p(\Sigma_R)}), \tag{5.32}$$

*where $c_1$ is a constant having the same dependence as above.*

To prove Corollary 5.1 let $d = R - R'$ and denote by $S_{x^0,r}$ the sphere $|x - x^0| < r$. Apply Lemma 5.1 to $u$ in $\bar{S}_{x^0,d}$ after obvious translation of variable), $x^0 \in \bar{S}_{R'}$. From the lemma, combined with Lemma 3.3, it follows that there exist positive constants $r_0 < d$ and $c_0$, both $r_0$ and $c_0$ depend only on $n, m, p, \lambda, K, R$ and $d$, such that $u \in H_{1,L_p}(S_{x^0,r_0})$ and

$$\| D_i u \|_{L_p(S_{x^0,r_0})} \leq c_0 (C + \| u \|_{L_p(S_{R})}), \qquad i = 1, \dots, n. \tag{5.33}$$

Covering $\bar{S}_{R'}$ by a finite number of spheres $S_{x^0,r_0}$ $(x^0 \in \bar{S}_{R'})$, using (5.33), we conclude that $u \in H_{1,L_p}(S_{R'})$ and that, furthermore, the inequality (5.31) holds.

Corollary 5.2 follows similarly from Lemma 5.2, Lemma 3.3′ and Corollary 5.1. (6).

## 6. The basic regularity theorems.

We pass to the main regularity results in the framework of the $L_p$ theory. The first theorem dealing with interior regularity is the following

THEOREM 6.1. *Let $A$ be an elliptic operator of order $2m$* (7) *defined in $\bar{S}_R$, with coefficients satisfying Condition $\{j; K\}$, $j$ being an integer such that*

---

(6) It should be observed here that in the exceptional case $n = 2$ the operator $A'(x^0, D)$ satisfies the «roots condition» for every $x^0 \in \bar{\Sigma}_R$. This follows by a simple continuity argument from our assumption that this is true for $x^0 = 0$.

(7) The theorem also holds for the elliptic operators in two variables of odd order.

*$1 \leq j \leq 2m$. Let, further, $u$ be a function such that $u \in L_q^{\text{loc.}}(S_R)$ for some $q > 1$ and such that for all functions $\varphi \in C_0^\infty(S_R)$ the following inequality holds :*

$$(6.1) \qquad |(u, A\varphi)_{S_R}| \leq C \|\varphi\|_{2m-j, L_{p'}(S_R)},$$

*where $p'$ is some fixed number $> 1$ and $C$ is a constant.*

*Denote by $p$ the exponent conjugate to $p'$: $\frac{1}{p} + \frac{1}{p'} = 1$. Then, $u \in H_{j,L_p}^{\text{loc.}}(S_R)$. Moreover, if $0 < R' < R$ and $R_1 = (R + R')/2$, then*

$$(6.2) \qquad \|u\|_{j,L_p(S_{R'})} \leq c_1 (C + \|u\|_{L_p(S_{R_1})}).$$

*where $c_1$ is a constant depending only on $n, m, p, \lambda, K, R$ and $R'$.*

*Proof*: Assume first that $j = 1$ and that $q \geq p$, so that we also have $u \in L_p^{\text{loc.}}(S_R)$. In this case the theorem follows from Corollary 5.1 applied to $u$ in $S_{R_1}$.

Next, let $j = 1$ but $1 < q < p$. To prove the theorem in this case it will suffice to show that actually $u \in L_p^{\text{loc.}}(S_R)$, thus reducing the proof to the case just established.

Now, let $q'$ be the exponent conjugate to $q$. Since $q' > p'$ it follows from (6.1) that we also have

$$|(u, A\varphi)_{S_R}| \leq \text{Const.} \|\varphi\|_{2m-1, L_{q'}(S_R)},$$

for all functions $\varphi \in C_0^\infty(S_R)$. Hence, by the result just established ($p$ replaced by $q$) we conclude that $u \in H_{1,L_q}^{\text{loc.}}(S_R)$. Invoking Sobolev's inequalities (Theorem 3.2) it follows that $u \in L_p^{\text{loc.}}(S_R)$ if either $q \geq n$ or $q < n$ but $q_1 = qn/(n-q) \geq p$. On the other hand if $q < n$ and $q_1 < p$ Sobolev's inequalities give only that $u \in L_{q_1}^{\text{loc.}}(S_R)$. In this case (noting that $q_1 > q$) we repeat the same argument with $q$ replaced by $q_1$; either arriving at the desired result $u \in L_p^{\text{loc.}}(S_R)$, or at least proving that $u \in L_{q_2}^{\text{loc.}}(S_R)$ with $q_2 > q_1$. Carrying on in this manner we obtain after a finite number of steps that $u \in L_p^{\text{loc.}}(S_R)$. This yields the theorem for $j = 1$.

To prove the theorem for $j \geq 2$ we use induction — supposing the theorem is true for $j - 1$ $(1 \leq j - 1 < 2m)$ we shall prove it for $j$.

We first observe that without loss of generality we may assume that $A$ contains no terms of order $\leq 2m - j$:

$$A = \sum_{2m-j < |\alpha| \leq 2m} a_\alpha D^\alpha .$$

(In the general case one can omit from $A$ all terms of order $\leq 2m - j$, the resulting operator will still satisfy an inequality of the form (6.1) in $S_{R_1}$ with $C$ replaced by $C + \| u \|_{L_p(S_{R_1})}$). Put $b_{\alpha,i} = D_i a_\alpha$ and define

$$B_i = \sum_{2m-j<|\alpha|\leq 2m} b_{\alpha,i} D^\alpha .$$

From the induction assumption it follows that $u \in H^{\text{loc.}}_{j-1,L_p}(S_R)$, and in particular that $D_i u \in L_p^{\text{loc.}}(S_R)$. Set $R_0 = (R' + R_1)/2$. For $\varphi \in C_0^\infty(S_{R_0})$ we have:

$$(6.3) \qquad (D_i u, A\varphi)_{S_{R_0}} = (u, D_i A \varphi)_{S_{R_0}} = (u, A D_i \varphi)_{S_{R_0}} + (u, B_i \varphi)_{S_{R_0}} .$$

Now, from (6.1) we get

$$(6.4) \qquad |(u, A D_i \varphi)_{S_{R_0}}| \leq C \| D_i \varphi \|_{2m-j, L_{p'}(S_{R_0})} \leq C \| \varphi \|_{2m-(j-1), L_{p'}(S_{R_0})} .$$

Also, since

$$(u, B_i \varphi)_{S_{R_0}} = \sum_{2m-j<|\alpha|\leq 2m} (u, b_{\alpha,i} D^\alpha \varphi)_{S_{R_0}} = (-1)^{j-1} \sum (D^\beta (\overline{b}_{\alpha,i} u), D^\gamma \varphi)_{S_{R_0}},$$

where $D^\beta D^\gamma = D^\alpha$ with $|\beta| = j-1$, $|\gamma| \leq 2m - (j-1)$, we find readily that

$$(6.5) \qquad |(u, B_i \varphi)_{S_{R_0}}| \leq \mu K \| u \|_{j-1, L_p(S_{R_0})} \| \varphi \|_{2m-(j-1), L_{p'}(S_{R_0})}$$

where $\mu$ is a constant depending only on $n$ and $m$.

Combining (6.3), (6.4) and (6.5) we obtain the inequality

$$(6.6) \qquad |(D_i u, A \varphi)_{S_{R_0}}| \leq C_1 \| \varphi \|_{2m-(j-1), L_{p'}(S_{R_0})}$$

which holds for all functions $\varphi \in C_0^\infty(S_{R_0})$, $i = 1, \ldots, n$, and where

$$(6.6') \qquad C_1 = C + \mu K \| u \|_{j-1, L_p(S_{R_0})} .$$

The inequality (6.6) shows that the derivatives $D_i u$ satisfy the conditions of the theorem in $S_{R_0}$ with $j$ replaced by $j-1$. Hence, using the induction assumption, we conclude that $D_i u \in H^{\text{loc.}}_{j-1,L_p}(S_{R_0})$ (and consequently $D_i u \in H^{\text{loc.}}_{j-1,L_p}(S_R)$). Also, denoting by $c_k$ constants which depend only on $n, m, p, \lambda, K, R$ and $R'$, we have:

$$(6.7) \qquad \| D_i u \|_{j-1, L_p(S_{R'})} \leq c_2 (C + \| u \|_{j-1, L_p(S_{R_0})}) .$$

Thus (using Lemma 3.2) we infer that $u \in H^{\text{loc.}}_{j,L_p}(S_R)$ and

$$(6.8) \qquad \| u \|_{j,L_p(S_{R'})} \leq c_3(C + \| u \|_{j-1,L_p(S_{R_0})}) .$$

Finally, noting that by the induction assumption we also have:

$$(6.8)' \qquad \| u \|_{j-1,L_p(S_{R_0})} \leq c_4 (C + \| u \|_{L_p(S_{R_1})}) ,$$

we derive from (6.8) and (6.8)′ the desired inequality (6.2). This establishes the theorem.

We now pass to the more refined result yielding regularity at the boundary. We shall first deal with functions defined in the hemisphere $\Sigma_R$, and with the regularity of such functions near the flat part of the boundary.

THEOREM 6.2. *Let $A(x, D)$ be an elliptic operator of order $2m$ defined in $\overline{\Sigma}_R$, with coefficients satisfying Condition $\{j ; K\}$, $j$ being an integer such that $1 \leq j \leq 2m$. If $n = 2$ assume in addition that $A'(x^0, D)$ satisfies the « roots condition » for every fixed $x^0 \in \overline{\Sigma}_R$.*

*Let, further, $u$ be a function belonging to $L_q(\Sigma_R)$ for some $q > 1$ such that*

$$(6.9) \qquad |(u , A\varphi)_{\Sigma_R}| \leq C \| \varphi \|_{2m-j,L_{p'}(\Sigma_R)} ,$$

*for all functions $\varphi \in C^\infty(\overline{\Sigma}_R)$ satisfying the boundary conditions:*

$$(6.9)' \qquad \begin{cases} D_n^k \varphi = \quad \text{on} \quad \partial_1 \Sigma_R \quad \text{for} \quad k = 0 , \dots , m-1 , \\ \varphi \equiv 0 \quad \text{in some neighborhood of} \ \partial_2 \Sigma_R , \end{cases}$$

*where $p'$ is some fixed number $> 1$ and $C$ is a constant.*

*Denote by $p$ the exponent conjugate to $p'$. Then, $u \in H_{j,L_p}(\Sigma_{R'})$ for every $R' < R$. Moreover, setting $R_1 = (R + R')/2$, we have:*

$$(6.10) \qquad \| u \|_{j,L_p(\Sigma_{R'})} \leq c_1 (C + \| u \|_{L_p(\Sigma_{R_1})})$$

*where $c_1$ is a constant depending only on $n , m , p , \lambda , K , R$* and *$R'$.*

The proof of the theorem depends on Corollary 5.2 and the following lemma.

LEMMA 6.1. *Let $u \in L_p(\Sigma_R)$, $p > 1$. Suppose that the distribution derivatives $D_i u$ for $i = 1 , \dots , n-1$ are functions belonging to $L_p(\Sigma_R)$. Suppose, moreover, that there exist an integer $l > 0$ and a constant $C_1$ such that*

$$(6.11) \qquad |(u , D_n^l \varphi)_{\Sigma_R}| \leq C_1 \| \varphi \|_{l-1,L_{p'}(\Sigma_R)}$$

*for all functions* $\varphi \in C_0^\infty(\Sigma_R)$ $\left(\frac{1}{p}+\frac{1}{p'}=1\right)$. *Then, the (distribution) derivative* $D_n u$ *is also a function belonging to* $L_p(\Sigma_{R'})$ *for every* $R' < R$, *and*

$$\| D_n u \|_{L_p(\Sigma_{R'})} \leq c \left( C_1 + \sum_{i=1}^{n-1} \| D_i u \|_{L_p(\Sigma_R)} + \| u \|_{L_p(\Sigma_R)} \right) \tag{6.12}$$

*where* $c$ *is a constant depending only on* $n, l, p, R$ and $R'$.

The lemma in the special case $p = 2$ is due to Lions (see [19]). For general $p$ (and also its analogue for Hölder classes of functions) the lemma is given (essentially) in Agmon-Douglis-Nirenberg [3] where, however, instead of (6.11) it is assumed that

$$D_n^l u = \sum_{|\alpha| \leq l-1} D^\alpha f_\alpha, \tag{6.13}$$

where $f_\alpha$ are certain functions belonging to $L_p(\Sigma_R)$ and (6.13) is understood in the weak (distribution) sense. Clearly, (6.13) implies an inequality of the form (6.11). The converse implication is also true (see [19]). For the sake of completeness we shall furnish in the sequel a variant proof of Lemma 6.1 which is not using the representation formula (6.13).

We shall now give the

*Proof of Theorem* 6.2. By the interior regularity result of Theorem 6.1 we know already that $u \in H^{\text{loc.}}_{j,L_p}(\Sigma_R)$. In the following we shall assume that the operator $A$ contains no terms of order $\leq 2m - j$. This (as in the proof of Theorem 6.1) entails no loss of generality. Given $0 < R' < R$, we set $R_1 = (R' + R)/2$ and $R_0 = (R' + R_1)/2$. We also denote by $c_k$ constants which depended only on $n, m, p, \lambda, K, R$ and $R'$.

To prove the theorem suppose first that $j = 1$ and that $q \geq p$ so that, in particular, $u \in L_p(\Sigma_R)$. Since in this case $u$ satisfies the conditions of Lemma 5.2 in $S_{R_1}$, applying Corollary 5.2, we conclude that

$$D_i u \in L_p(\Sigma_{R_0}) \qquad \text{for} \qquad i = 1, \dots, n-1,$$

and that

$$\sum_{i=1}^{n-1} \| D_i u \|_{L_p(\Sigma_{R_0})} \leq c_2 (C + \| u \|_{L_p(\Sigma_{R_1})}). \tag{6.14}$$

To complete the proof in this case we need only to show that $D_n u \in L_p(\Sigma_{R'})$ and that

$$\| D_n u \|_{L_p(\Sigma_{R'})} \leq c_3 (C + \| u \|_{L_p(\Sigma_{R_1})}). \tag{6.14'}$$

To this end write $A$ in the form:

$$A = aD_n^{2m} + \sum_{i=1}^{n-1} \sum_{|\alpha|=2m-1} a_{\alpha,i} D_i D^\alpha, \tag{6.15}$$

where $a_{\alpha,i}$ is either a coefficient in $A$ or zero. Let $\varphi \in C_0^\infty(\Sigma_{R_0})$. We have:

$$\begin{aligned}(\bar{a}\, u, D_n^{2m} \varphi)_{\Sigma_{R_0}} &= (u, a D_n^{2m} \varphi)_{\Sigma_{R_0}} \\ &= (u, A\varphi)_{\Sigma_{R_0}} + \sum_{i=1}^{n-1} \sum_{|\alpha|=2m-1} (D_i(\bar{a}_{\alpha,i} u), D^\alpha \varphi)_{\Sigma_{R_0}}.\end{aligned} \tag{6.16}$$

Combining (6.16), (6.14) and (6.9) we infer

$$|(\bar{a}\, u, D_n^{2m} \varphi)_{\Sigma_{R_0}}| \leq c_4 (C + \| u \|_{L_p(\Sigma_{R_1})}) \| \varphi \|_{2m-1, L_{p'}(\Sigma_{R_0})} \tag{6.17}$$

for all functions $\varphi \in C_0^\infty(\Sigma_{R_0})$.

Applying now Lemma 6.1 to the function $\bar{a}\, u$ in $\Sigma_{R_0}$ we conclude that $D_n(\bar{a}\, u) \in L_p(\Sigma_{R'})$ and that

$$\| D_n(\bar{a}\, u) \|_{L_p(\Sigma_{R'})} \leq c_5 (C + \| u \|_{L_p(\Sigma_{R_0})} + \sum_{i=1}^{n-1} \| D_i(\bar{a}\, u) \|_{L_p(\Sigma_{R_0})}). \tag{6.18}$$

Combining (6.18) and (6.14) (using $\lambda^{-1} \leq |a| \leq \lambda$, $|D_i a| \leq K$) we conclude that $D_n u \in L_p(\Sigma_{R'})$ and that (6.14)′ holds. This yields the theorem in the case considered.

Next, suppose that $j = 1$ but that $1 < q < p$. We shall reduce this case to the preceding one by showing that actually $u \in L_p(\Sigma_{R'})$ for every $R' < R$. We proceed as in the proof of Theorem 6.1. By the above argument ($p$ replaced by $q$) we have $u \in H_{1,L_q}(\Sigma_{R'})$. Hence, using Sobolev's inequalities we conclude that $u \in L_p(\Sigma_{R'})$ if $q \geq n$, or if $q < n$ but $q_1 = q\, n/(n-q) \geq p$. On the other hand, if $q < n$ and $q < q_1 < p$ we conclude that $u \in L_{q_1}(\Sigma_R)$. Repeating in the last case the same argument with $q$ replaced by $q_1$ etc., we conclude after a finite number of steps that $u \in L_p(\Sigma_{R'})$ for every $R' < R$. This completes the proof of the theorem for $j = 1$.

Finally, for $j \geq 2$ we use induction — assuming the theorem is true for $j-1$ $(1 \leq j-1 < 2m)$ we establish it for $j$. From the induction assumption we note that $u \in H_{j-1,L_p}(\Sigma_{R'})$ so that in particular $D_i u \in L_p(\Sigma_{R'})$ for every $R' < R$.

Consider a derivative $D_i u$ with $i \neq n$. Let $\varphi$ be a function belonging to $C^\infty(\bar{\Sigma}_R)$ with support in $\bar{\Sigma}_{R_0}$, such that

$$D_n^k \varphi = 0 \quad \text{for} \quad x_n = 0\,, \qquad k = 0\,, \dots, m-1\,. \tag{6.19}$$

We now proceed to estimate $(D_i u\,, A\,\varphi)_{\Sigma_{R_0}}$ in exactly the same manner as in the proof of Theorem 6.1. Using the fact that $D_i\,\varphi\,,\ i \neq n$, satisfies (6.9)′, we need only to rewrite relations (6.3) to (6.6), replacing everywhere the sphere $S_{R_0}$ by the hemisphere $\Sigma_{R_0}$. Rewriting the final inequality (6.6) we thus conclude that

$$|(D_i u, A\,\varphi)_{\Sigma_{R_0}}| \leq C_1 \|\varphi\|_{2m-(j-1), L_{p'}(\Sigma_{R_0})} \tag{6.20}$$

for $i = , \dots, n-1$ and all functions $\varphi \in C^\infty(\bar{\Sigma}_R)$ with support in $\bar{\Sigma}_{R_0}$, satisfying the boundary conditions (6.19). Here $C_1$ is the constant (6.6)′.

Hence the derivatives $D_i u\,(i \neq n)$ satisfy the conditions of Theorem 6.2 in $\Sigma_{R_0}$ with $j$ replaced by $j-1$. Applying the induction, setting $R'' = (R' + R_0)/2$, we infer that

$$\sum_{i=1}^{n-1} \|D_i u\|_{j-1, L_p(\Sigma_{R''})} \leq c_6 (C + \|u\|_{j-1, L_p(\Sigma_{R_0})}) \tag{6.21}$$

$$\leq c_7 (C + \|u\|_{L_p(\Sigma_{R_1})})\,.$$

From (6.21) and from the validity of the theorem for $j-1$ it follows that all derivatives $D^\alpha u$ such that $0 \leq |\alpha| \leq j\,,\ \alpha \neq (0\,, \dots, 0\,, j)$ belong to $L_p(\Sigma_{R'})$ and satisfy

$$\|D^\alpha u\|_{L_p(\Sigma_{R'})} \leq c_8 (C + \|u\|_{L_p(\Sigma_{R_1})})\,. \tag{6.22}$$

To complete the proof of the theorem we need only to show that $D_n^j u \in L_p(\Sigma_{R'})$ and satisfies (6.22). This we do again with the aid of Lemma 6.1. Write $A$ in the form:

$$A = a D_n^{2m} + \sum_{i=1}^{n-1} \sum_{|\alpha| = 2m-1} a_{\alpha,i} D_i D^\alpha + \sum_{2m-j \leq |\alpha| \leq 2m-1} a_\alpha D^\alpha\,. \tag{6.23}$$

Let $\varphi \in C_0^\infty(\Sigma_{R''})$. Using integration by parts, we have:

$$(-1)^{j-1} (D_n^{j-1} (\bar{a} u)\,, D_n^{2m-j+1} \varphi)_{\Sigma_{R''}} = (u\,, a D_n^{2m} \varphi)_{\Sigma_{R''}} \tag{6.24}$$

$$= (u\,, A\varphi)_{\Sigma_{R''}} + \sum_{i=1}^{n-1} \sum_{|\alpha| = 2m-1} (D_i (\bar{a}_{\alpha,i} u)\,, D^\alpha \varphi)_{\Sigma_{R''}} -$$

$$- \sum_{2m-j \leq |\alpha| \leq 2m-1} (\bar{a}_\alpha u\,, D^\alpha \varphi)_{\Sigma_{R''}}\,.$$

Consider a typical term in the first sum on the right of (6.24). Writing $D^\alpha = D^\beta D^\gamma$, with $|\beta| = j-1$ and $|\gamma| = 2m-j$, we find readily that

$$(6.24)' \qquad |(D_i(\bar{a}_{\alpha,i} u), D^\alpha \varphi)_{\Sigma_{R''}}| = |(D^\beta D_i(a_{\alpha,i} u), D^\gamma \varphi)_{\Sigma_{R''}}|$$

$$\leq c_9 (\| D_i u \|_{j-1, L_p(\Sigma_{R''})} + \| u \|_{L_p(\Sigma_{R''})}) \| \varphi \|_{2m-j, L_{p'}(\Sigma_{R''})}.$$

Similarly, for a typical term in the last sum on the right of (6.24) we find (integrating $|\alpha| - (2m-j)$ times by parts):

$$(6.24)'' \qquad |(\bar{a}_\alpha u, D^\alpha \varphi)_{\Sigma_{R''}}| \leq c_{10} \| u \|_{j-1, L_p(\Sigma_{R''})} \| \varphi \|_{2m-j, L_{p'}(\Sigma_{R''})}.$$

Combining (6.24), (6.24)′, (6.24)″ and (6.21) we thus get for all functions $\varphi \in C_0^\infty(\Sigma_{R''})$:

$$(6.25) \qquad |D_n^{j-1}(\bar{a}u), D_n^{2m-j+1} \varphi)_{\Sigma_{R''}}|$$

$$\leq c_{11}\left(C + \sum_{i=1}^{n-1} \| D_i u \|_{j-1, L_p(\Sigma_{R''})} + \| u \|_{j-1, L_p(\Sigma_{R''})}\right) \| \varphi \|_{2m-1, L_{p'}(\Sigma_{R''})} \leq$$

$$\leq c_{12} (C + \| u \|_{L_p(\Sigma_{R_1})}) \| \varphi \|_{2m-j, L_{p'}(\Sigma_{R''})}.$$

Applying, now, Lemma 6.1 to the function $D_n^{j-1}(\bar{a}u)$ in $\Sigma_{R''}$ with $l = 2m - j + 1$, we conclude that $D_n^j(\bar{a}u) \in L_p(\Sigma_{R'})$ and that

$$(6.26) \qquad \| D_n^j (\bar{a}u) \|_{L_p(\Sigma_{R'})} \leq c_{13} (C + \| u \|_{L_p(\Sigma_{R_1})}).$$

Finally, from (6.26) and (6.22) it follows that $D_n^j u \in L_p(\Sigma_{R'})$ and that $D_n^j u$ satisfies an inequality of the form (6.22). This completes the proof of the theorem.

We shall conclude this section presenting a

*Proof of Lemma* 6.1. We first note that by an obvious approximation argument the inequality

$$(6.27) \qquad |(u, D_n^l \varphi)_{\Sigma_R}| \leq C_1 \| \varphi \|_{l-1, L_{p'}(\Sigma_R)}$$

holds not only for $\varphi \in C_0^\infty(\Sigma_R)$, but also for all functions $\varphi \in C^l(\Sigma_R)$ with compact support in $\Sigma_R$. Moreover, we claim that (6.27) holds for all functions $\varphi \in C^l(\bar{\Sigma}_R)$ satisying the boundary conditions:

$$(6.27)' \qquad \begin{cases} D_n^j \varphi = 0 & \text{on} \quad \partial_1 \Sigma_R, \quad j = 0, \ldots, l, \\ \quad \varphi \equiv 0 & \text{in a neighborhood of } \partial_2 \Sigma_R. \end{cases}$$

Indeed, if $\varphi$ is such a function and $\varepsilon > 0$, define:

$$\begin{cases} \varphi_\varepsilon(x) = \varphi(x - \varepsilon) & \text{for} \quad x \in \Sigma_R, \quad x \geq \varepsilon, \\ \varphi_\varepsilon(x) \equiv 0 & \text{for} \quad x \in \Sigma_R, \quad x < \varepsilon. \end{cases}$$

If $\varepsilon$ is sufficiently small then $\varphi_\varepsilon$ will be a $C^l$ function with compact support in $\Sigma_R$. Hence, by the previous remark, (6.27) holds for $\varphi_\varepsilon$. Letting $\varepsilon \to 0$ we establish the same for $\varphi$.

Now, define $u$ as zero in $x_n > 0$, $|x| \geq R$. Then, extend $u$ into the half-space $x_n < 0$, putting

$$u(x', x_n) = \sum_{j=1}^{2l+1} \lambda_j u\left(x', -\frac{x_n}{j}\right) \quad \text{for} \quad x_n < 0, \tag{6.28}$$

$x' = (x_1, \dots, x_{n-1})$, where the constants $\lambda_j$ are chosen so that

$$\sum_{j=1}^{2l+1} \lambda_j \left(-\frac{1}{j}\right)^k = 1 \quad \text{for} \quad k = -1, 0, 1, \dots, 2l-1. \tag{6.28$'$}$$

Clearly $u \in L_p(S_R)$ and

$$\| u \|_{L_p(S_R)} \leq \gamma_1 \| u \|_{L_p(\Sigma_R)}. \tag{6.29}$$

where here and in the following $\gamma_1, \dots, \gamma_5$, denote constants depending only on $l$. Also, the distribution derivatives $D_i u$, for $i \neq n$ are functions belonging to $L_p(S_R)$ such that

$$\| D_i u \|_{L_p(S_R)} \leq \gamma_2 \| D_i u \|_{L_p(\Sigma_R)}. \tag{6.29$'$}$$

Let, now, $\chi$ be an arbitrary function of $C_0^\infty(S_R)$, extended as zero outside $S_R$. Write

$$(u, D_n^{2l} \chi)_{S_R} = \int_{x_n > 0} u \overline{D_n^{2l} \chi} \, dx + \int_{x_n < 0} u \overline{D_n^{2l} \chi} \, dx. \tag{6.30}$$

Using (6.28) to transform the last integral in (6.30), we find that

$$(u, D_n^{2l} \chi)_{S_R} = \int_{x_n > 0} u \overline{D_n^{2l} \chi^*} \, dx, \tag{6.30$'$}$$

where

$$\chi^*(x', x_n) = \chi(x', x_n) + \sum_{j=1}^{2l+1} \lambda_j \left(\frac{1}{j}\right)^{2l-1} \chi(x', -jx_n). \tag{6.31}$$

It is readily seen from (6.31) and (6.28)′ that $\chi^*$ is a $C^\infty$ function with support in $S_R$ such that

$$D_n^j \chi^* = 0 \quad \text{for} \quad x_n = 0, \quad j = 0, \dots, 2l. \tag{6.31$'$}$$

Putting $\varphi = D_n^l \chi^*$, we write (6.30′) in the form:

$$(u, D_n^{2l} \chi)_{S_R} = (u, D_n^l \varphi)_{\Sigma_R}. \tag{6.32}$$

Since, now, $\varphi \in C^\infty(\bar{\Sigma}_R)$ and satisfies the boundary conditions (6.27)′ it follows from the preceding that it satisfies (6.27). In terms of $\chi$ (using (6.31), (6.32)) we have

$$|(u, D_n^{2l} \chi)_{S_R}| \leq \gamma_3 C_1 \| \chi \|_{2l-1, L_{p'}(\Sigma_R)}. \tag{6.33}$$

Next, for $i \neq n$, we have:

$$(u, D_i^{2l} \chi)_{S_R} = -(D_i u, D_i^{2l-1} \chi)_{S_R},$$

from which, using (6.29)′, we obtain that

$$|(u, D_i^{2l} \chi)_{S_R}| \leq \gamma_4 \| D_i u \|_{L_p(\Sigma_R)} \| \chi \|_{2l-1, L_{p'}(S_R)}. \tag{6.33$'$}$$

Let $A = D_1^{2l} + \dots + D_n^{2l}$. Combining (6.33) and (6.33)′ we conclude that

$$|(u, A\chi)_{S_R}| \leq \gamma_5 \left(C_1 + \sum_{i=1}^{n-1} \| D_i u \|_{L_p(\Sigma_R)}\right) \| \chi \|_{2l-1, L_{p'}(\Sigma_R)} \tag{6.34}$$

for all functions $\chi \in C_0^\infty(S_R)$.

Since $A$ is elliptic, the inequality (6.34) allows us to apply Theorem 6.1 (Corollary 5.1) to $u$ in $S_R$. We conclude that $u \in H_{1, L_p}^{\text{loc.}}(S_R)$ and that for every $R' < R$ (using (6.29))

$$\| D_n u \|_{L_p(S_{R'})} \leq c \left( C + \sum_{i=1}^{n-1} \| D_i u \|_{L_p(\Sigma_R)} + \| u \|_{L_p(\Sigma_R)} \right),$$

where $c$ is a constant depending only on $n, l, p, R$ and $R'$. This establishes the lemma.

## 7. The interior regularity in $L_p$ of weak solutions of elliptic equations and overdetermined systems.

Let $\{A_i(x, D)\}_{i=1}^{N}$ be a system of $N$ linear differential operators of respective order $m_i$:

$$A_i(x, D) = \sum_{|\alpha| \leq m_i} a_\alpha^i(x) D^\alpha \qquad (i = 1, \dots, N), \tag{7.1}$$

defined in a closed bounded domain $\overline{G}$. We shall say that the system $\{A_i\}$ is elliptic in $\overline{G}$ if there exists no real vector $\xi \neq 0$ and a point $x \in \overline{G}$, such that

$$\sum_{|\alpha| = m_i} a_\alpha^i(x) \xi^\alpha = 0 \quad \text{for} \quad i = 1, \dots, N \quad (\xi^\alpha = \xi_1^{\alpha_1} \dots \xi_n^{\alpha_n}). \tag{7.2}$$

If the leading coefficients of $A_i$ are continuous in $\overline{G}$, ellipticity implies that there exists a constant $\lambda \geq 1$ such that

$$\frac{1}{\lambda^2} \leq \sum_{i=1}^{N} \Big| \sum_{|\alpha| = m_i} a_\alpha^i(x) \xi^\alpha \Big|^2 \leq \lambda^2, \tag{7.2$'$}$$

for all real unit-vectors $\xi$ and $x \in \overline{G}$. We term such a constant an ellipticity constant of the system.

For an overdetermined system of operators having the same order the above definition of ellipticity coincides with given by Schwartz [29] (see also Hörmander [15]). We point out, however, that we are not imposing the restriction that the operators $A_i$ be of the same order.

In the following $\{A_i\}_{i=1}^{N}$ will denote either an elliptic operator ($N = 1$) or an elliptic overdetermined system ($N \geq 2$) defined in $\overline{G}$ and given by (7.1). The formally adjoint $A_i^*$ of $A_i$ is the operator

$$A_i^*(x, D) u = \sum_{|\alpha| \leq m_i} (-1)^{|\alpha|} D^\alpha \left(\overline{a_\alpha^i(x)}\, u\right). \tag{7.3}$$

It is a differential operator in the ordinary sense if $a_\alpha^i \in C^{|\alpha|}(\overline{G})$. Clearly the system $\{A_i^*\}$ will also be elliptic.

We shall consider a weak solution $u$ of the adjoint system

$$A_i^* u = f_i, \quad i = 1, \dots, N, \tag{7.4}$$

in the sense that

$$(u, A_i \varphi)_G = (f_i, \varphi)_G, \; i = 1, \dots, N, \tag{7.5}$$

for all functions $\varphi \in C_0^\infty(G)$. Note that (7.5) has a sense when the coefficients of $A_i$ are merely measurable bounded functions.

The main interior $L_p$ regularity result for such weak solutions is the following

THEOREM 7.1. *Let $u$ be a function belonging to $L_q^{\text{loc.}}(G)$ for some $q > 1$. Suppose that $u$ satisfies (7.5) where $f_i$ $(i = 1, \dots, N)$ are given functions belonging to $L_p^{\text{loc.}}(G)$, $p > 1$, and where $\{A_i\}_{i=1}^N$ $(N \geq 1)$ is the elliptic system introduced above. Assume also that the coefficients of $A_i$ satisfy Condition $\{l; K\}$ in $\overline{G}$, $l$ being some positive integer* (8), *and put $j = \min(l, m_1, \dots, m_N)$. Then, $u \in H_{j,L_p}^{\text{loc.}}(G)$. Moreover, if $G_0$, $G_1$ are any two subdomains $G_0$, $G_1$, such that $\overline{G}_0 \subset G_1 \subset \overline{G}_1 \subset G$, then*

$$\| u \|_{j,L_p(G_0)} \leq c \left( \sum_{i=1}^N \| f_i \|_{L_p(G_1)} + \| u \|_{L_p(G_1)} \right), \tag{7.6}$$

*where $c$ is a constant depending only on $n$, $\max m_i$, $p$, $N$, $K$, the ellipticity constant $\lambda$ and the domains.*

*Proof*: Put $m_0 = \min m_i$, $m = \max m_i$, and let $d$ be the distance between $\partial G_0$ and $\partial G_1$. Denote by $\overline{A}_i$ the differential operator with coefficients complex conjugate to those of $A_i$. Given a point $x^0 \in \overline{G}_0$, define:

$$A_{x^0}(x, D) = \sum_{i=1}^N A_i(x, D)\, \overline{A_i}(x^0, D)\, \Delta^{m-m_i}$$

where $\Delta$ is the Laplacean. $A_{x^0}$ is a linear differential operator of order $2m$ with coefficients satisfying Condition $\{l; c_0 K\}$ in $\overline{G}$, $c_0$ being some constant depending only on $n$, $m$, and $N$. Also, $A_{x^0}$ is elliptic at $x^0$ and consequently, by continuity, is elliptic in some neighborhood of $x^0$. More precisely, since the coefficients of the leading part $A'_{x^0}$ possess first derivatives bounded by $c_0 K$, it is readily seen that there exists a positive number $\varrho \leq d$, $\varrho$ depending only on $n$, $m$, $N$, $K$, $\lambda$ and $d$, such that

$$\frac{1}{2\lambda^2} | \xi |^{2m} \leq | A'_{x^0}(x, \xi) | \leq 2\lambda^2 | \xi |^{2m}$$

for $| x - x^0 | \leq \varrho$ and all real vectors $\xi$. Thus, denoting by $S_{x^0,r}$ the sphere $| x - x^0 | < r$, $A_{x^0}$ is elliptic in $\overline{S}_{x^0,\varrho}$ and $2\lambda^2$ can serve as its ellipticity constant.

---

(8) Condition $\{l; K\}$ for the coefficients of $A_i$ is defined as in § 5 (Def. 5.1) except that $2m$ should be replaced by the order $m_i$ of $A_i$.

Now, let $\varphi \in C_0^\infty(S_{x^0,\varrho})$. By (7.5) we have:

$$(u, A_i(x, D)\bar{A}_i(x^0, D)\Delta^{m-m_i}\varphi)_{S_{x^0,\varrho}} = (f_i, \bar{A}_i(x^0, D)\Delta^{m-m_i}\varphi)_{S_{x^0,\varrho}},$$

which after summation yields:

$$(7.7)\qquad (u, A_{x^0}\varphi)_{S_{x^0,\varrho}} = \sum_{i=1}^{N}(f_i, \bar{A}_i(x^0, D)\Delta^{m-m_i}\varphi)_{S_{x^0,\varrho}}$$

It follows from (7.7) that

$$(7.8)\qquad |(u, A_{x^0}\varphi)_{S_{x^0,\varrho}}| \leq C\sum_{i=1}^{N}\|f_i\|_{L_p(S_{x^0,\varrho})}\|\varphi\|_{2m-m_0, L_{p'}(S_{x^0,\varrho})}$$

where $C$ is a constant depending only on $n, m, N$, and $K$.

The conclusion of the theorem follows now immediately from (7.8) and Theorem 6.1 applied to $u$ in $S_{x^0,\varrho}$ (elliptic operator $A_{x^0}$), using a finite covering of $\bar{G}_0$ by spheres $S_{x^i,\varrho/2}$ ($x^i \in \bar{G}_0$).

The following is an easy consequence, and at the same time a generalization, of Theorem 7.1.

Theorem 7.1'. *Suppose that the conditions of Theorem 7.1 hold and that in addition* $f_i \in H^{\mathrm{loc.}}_{k_i, L_p}(G)$, $k_i \geq 0$. *Set* $k = \min(l, k_1 + m_1, \ldots, k_N + m_N)$. *Then* $u \in H^{\mathrm{loc.}}_{k, L_p}(G)$, *and for any two subdomains* $G_0, G_1$ *such that* $\bar{G}_0 \subset G_1 \subset \subset \bar{G}_1 \subset G$ *the following inequality holds:*

$$(7.9)\qquad \|u\|_{k, L_p(G_0)} \leq c\left(\sum_{i=1}^{N}\|f_i\|_{k_i, L_p(G_1)} + \|u\|_{L_p(G_1)}\right),$$

*where c is a constant depending only on* $n$, $\max(m_i + k_i)$, $N, p, K, \lambda$ *and the domains.*

*Proof*: The special case $k_1 = \ldots = k_N = 0$ is Theorem 7.1. In the general case put:

$$A_{i,\alpha}(x, D) = A_i(x, D)D^\alpha \quad \text{and} \quad f_{i,\alpha} = (-1)^{k_i}D^\alpha f_i,$$

for $|\alpha| = k_i$, $i = 1, \ldots, N$. Integrating by parts we deduce from (7.5) that

$$(7.10)\qquad (u, A_{i,\alpha}\varphi)_G = (f_{i,\alpha}, \varphi)_G$$

for $\varphi \in C_0^\infty(G)$, $|\alpha| = k_i$, $i = 1, \ldots, N$.

The conclusion of the theorem follows now from (7.10) and Theorem 7.1 applied to the function $u$, elliptic system $\{A_{i,\alpha}\}$ and the corresponding system of functions $\{f_{i,\alpha}\}$.

Suppose now that the conditions of Theorem 7.1′ hold with $k_i = m - m_i$ and $l = m$. It follows from the theorem that $u \in H^{\text{loc.}}_{m,L_p}(G)$. Using integration by parts it follows in a standard way from (7.5) that $u$ is a strong solution (in $H^{\text{loc.}}_{m,L_p}$) of the adjoint system (7.4). If, moreover, the conditions of Theorem 7.1′ hold with $k_i = m - m_i + j$, $k = m + j$, where $j > n/p$, then it follows from Sobolev's inequalities that $u \in C^m(G)$, $f_i \in C(G)$ (after correction on a null set) and that $u$ satisfies (7.4) in the classical sense. Finally, if the coefficients of the system and the $f_i$ are infinitely differentiable one obtains the well known result that $u$ is also infinitely differentiable (for overdetermined elliptic systems see, for instance, Schwartz [29]).

With the aid of Theorem 7.1 we establish now the following a priori estimates for a system of differential operators.

THEOREM 7.2. *Let* $\{A_i\}_{i=1}^{N}$ *be an elliptic system of differential operators of respective order* $m_i$ *defined in* $\overline{G}$. *Set* $m_0 = \min m_i$, *and suppose that the coefficients of* $A_i$ *satisfy Condition* $\{m_i; K\}$ *in* $\overline{G}$. *Let* $G_0$ *be a subdomain such that* $\overline{G}_0 \subset G$. *Then, for all functions* $u \in C_0^\infty(G_0)$:

$$\| u \|_{m_0, L_p(G_0)} \leq c \left( \sum_{i=1}^{N} \| A_i u \|_{L_p(G_0)} + \| u \|_{L_p(G_0)} \right) \tag{7.11}$$

*where* $c$ *is a constant independent of* $u$.

*Proof:* Put $A_i u = f_i$. Then, for every function $\varphi \in C_0^\infty(G)$:

$$(u, A_i^* \varphi)_G = (f_i, \varphi)_G, \quad i = 1, \ldots, N, \tag{7.12}$$

where $\{A_i^*\}$ is the formally adjoint system. The inequality (7.11) follows now from (7.12) and from Theorem 7.1 applied to $u$ in $G$, system $\{A_i^*\}$ and $l = j = m_0$.

The estimate (7.11) for a single elliptic operator was established by various authors (see, for instance, Nirenberg [24]). For $p = 2$ and $m_1 = m_2 = \ldots = m_N$ estimate follows from the more general Gårding's inequality [14]. For general $p$ the estimate (7.11) was (essentially) established in Agmon-Douglis-Nirenberg [3; Th. 15.1″] by a different method.

In the special case of an elliptic system of operators having the same order the smoothness assumptions imposed on the coefficients of $A_i$ in Theorem 7.2 could be relaxed considerably, namely, we have

THEOREM 7.2′. *Let* $\{A_i(x, D)\}_{i=1}^{N}$ *be an elliptic system of operators in* $\overline{G}$, *having the same order* $m$. *Suppose that the coefficients of highest order terms in* $A_i$ *are continuous, whereas the remaining coefficients are measurable and* *bounded in $\overline{G}$. Then, for all functions $u \in C_0^\infty(G)$ we have:*

$$\| u \|_{m, L_p(G)} \leq c \left( \sum_{i=1}^{N} \| A_i u \|_{L_p(G)} + \| u \|_{L_p(G)} \right), \tag{7.13}$$

*where $c$ is a constant independent of $u$.*

We sketch the proof. Using Lemma 3.4 we may assume without loss of generality that $A_i(x, D)$ contains no terms of order $< m$. Let $x^0$ be an arbitrary point of $\overline{G}$ and put $A_i^0 = A_i(x^0, D)$. By Theorem 7.2 the inequality (7.13) holds for the elliptic system with constant coefficients $\{A_i^0\}$. Hence, there exists a constant $c_0 > 0$ such that for all $u \in C_0^\infty(G)$ we can write

$$\| u \|_{m, L_p(G)} \leq c_0 \left( \sum_{i=1}^{N} \| A_i^0 u \|_{L_p(G)} + \| u \|_{L_p(G)} \right) \tag{7.14}$$

$$\leq c_0 \left( \sum_{i=1}^{N} \| A_i u \|_{L_p(G)} + \| u \|_{L_p(G)} \right) + c_0 \sum_{i=1}^{N} \| (A_i^0 - A_i) u \|_{L_p(G)}.$$

Using the continuity of the coefficients of $A_i$ it is readily seen that there exists a number $\varrho > 0$ (independent of $x^0$) such that if the support of $u$ is contained in the sphere $|x - x^0| < \varrho$, then the last term on the right of (7.14) is less than $\frac{1}{2} \| u \|_{m, L_p(G)}$. From this and (7.14) it follows that there exists a number $\delta > 0$ such that (7.13) holds for all functions $u \in C_0^\infty(G)$ which in addition possess support of diameter $< \delta$. Finally, one drops the restriction on the support of $u$ in a standard way by using a suitable partition of unity and using once more Lemma 3.4.

## 8. Regularity at the boundary.

We pass to the problem of regularity at the boundary in $L_p$ of weak solutions of the Dirichlet problem. We consider an elliptic operator $A$ of order $2m$ defined in $\overline{G}$:

$$A(x, D) = \sum_{|\alpha| \leq 2m} a_\alpha(x) D^\alpha. \tag{8.1}$$

If $n = 2$ we assume in addition that $A$ satisfies the roots condition in $\overline{G}$ (i. e. for every $x^0 \in \overline{G}$ the principal part $A'(x^0, D)$ satisfies the condition on the roots introduced in § 4). We denote by $C^l(\overline{G}; \{D^\alpha\}_{|\alpha| \leq m-1})$ $(m \leq l)$ the subclass of function $v \in C^l(\overline{G})$ satisfying the boundary conditions:

$$D^\alpha v = 0 \quad \text{on} \quad \partial G \quad \text{for} \quad 0 \leq |\alpha| \leq m - 1. \tag{8.2}$$

We also recall that $H_{l,L_p}(\overline{G}\,;\{D^\alpha\}_{|\alpha|\le m-1})$ denotes the subclass of functions $v \in H_{l,L_p}(G)$ satisying (8.2) in the generalized (trace) sense (see § 2).

We now state the basic

THEOREM 8.1. *Let $u$ be a function belonging to $L_q(G)$ for some $q>1$. Suppose that for all functions $v \in C^{2m}(\overline{G}\,;\{D^\alpha\}_{|\alpha|\le m-1})$ the following inequality holds:*

$$(8.3)\qquad |(u, Av)_G| \le C \| v \|_{2m-j, L_{p'}(G)},$$

*where $A$ is the elliptic operator* (8.1), *$j$ is a positive integer $\le 2m$, $p'>1$ and $C$ a constant. Suppose also that the coefficients of $A$ satisfy condition $\{j, K\}$ in $\overline{G}$ and that $G$ is of class $C^{2m}$. Then, $u \in H_{j,L_p}(G)\left(\dfrac{1}{p}+\dfrac{1}{p'}=1\right)$ and*

$$(8.4)\qquad \| u \|_{j,L_p(G)} \le c_1 (C + \| u \|_{L_p(G)}),$$

*where $c_1$ is a constant depending only on $n, m, p, K, \lambda$ (the ellipticity constant), and the domain.*

*Proof:* By an obvious covering argument it suffices to show that for every $x^0 \in \overline{G}$ there exists a neighborhood $\Omega^0$ in the relative topology of $\overline{G}$ such that $u \in H_{j,L_p}(\Omega^0)$, and such that $\| u \|_{j,L_p(\Omega^0)}$ is majorized by the right side of (8.4) with a constant $c_1$ depending in addition on $\Omega^0$. For a point $x^0$ in the interior this follows from Theorem 7.1, taking for $\Omega^0$ a sufficiently small sphere with center at $x^0$. Suppose that $x^0 \in \partial G$. In this case there exists a sufficiently small neighborhood $\Omega$ of $x^0$ in $\overline{G}$, and a measure preserving homeomorphism (9) of class $C^{2m}$: $x \to \tilde{x}$ which takes $\overline{\Omega}$ onto the hemisphere $\overline{\Sigma}_1$: $|\tilde{x}| \le 1$, $\tilde{x}_n \ge 0$. Let $\tilde{A}$ be the transformed elliptic operator under the mapping and put $\tilde{u}(\tilde{x}) = u(x(\tilde{x}))$ ($\tilde{A}$ and $\tilde{u}$ defined in $\overline{\Sigma}_1$). Let, further, $\tilde{v}$ be an arbitrary function belonging to $C^{2m}(\overline{\Sigma}_1\,;\{D^\alpha\}_{|\alpha|\le m-1})$ and vanishing in some neighborhood of $\partial_2\Sigma_1$ (the curved part of $\partial\Sigma_1$). Put $v(x) = \tilde{v}(\tilde{x}(x))$ and extend $v$ as zero in $\overline{G} - \overline{\Omega}$. It is readily seen that $v \in C^{2m}(\overline{G}\,;\{D^\alpha\}_{|\alpha|\le m-1})$. Using (8.3) we have:

$$(8.5)\qquad \begin{aligned} |(\tilde{u}, \tilde{A}\tilde{v})_{\Sigma_1}| &= |(u, A v)_G| \\ &\le C \| v \|_{2m-j, L_{p'}(G)} \le c_0\, C \| \tilde{v} \|_{2m-j, L_{p'}(\Sigma_1)} \end{aligned}$$

---

(9) One can take a mapping of the form: $\tilde{x}_1 = x_1, \ldots,$

$$\tilde{x}_{k-1} = x_{k-1}, \quad \tilde{x}_k = x_{k+1}, \ldots, \quad \tilde{x}_{n-1} = x_n,$$

$$\tilde{x}_n = x_k - f(x_1, \ldots, x_{k-1}, x_{k+1}, \ldots, x_n).$$

where $c_0$ depends only on the mapping. Applying now Theorem 6.2 to the function $\tilde{u}$ in $\Sigma_1$ we conclude that $\tilde{u} \in H_{j,L_p}(\Sigma_r)$ for every $r < 1$ and consequently that $u \in H_{j,L_p}(\Omega^0)$, $\bar{\Omega}^0$ being the image of $\bar{\Sigma}_r$ under the mapping. We also obtain by the same theorem the desired estimate. This establishes Theorem 8.1.

From Theorem 8.1 one deduces easily the regularity up to the boundary of weak solutions of the Dirichlet problem:

$$\text{(8.6)} \qquad \begin{cases} A^* u = f & \text{in} \quad G, \\ D^\alpha u = 0 & \text{on} \quad \partial G, \quad 0 \leq |\alpha| \leq m-1. \end{cases}$$

THEOREM 8.2. *Let $u$ be a function belonging to $L_q(G)$ for some $q > 1$. Suppose that $u$ is a weak solution of* (8.6) *in the sense that*

$$\text{(8.7)} \qquad (u, Av)_G = (f, v)_G$$

*for all functions $v \in \overset{\bullet}{C}{}^{2m}(\bar{G}; \{D^\alpha\}_{|\alpha| \leq m-1})$, where $A$ is the elliptic operator* (8.1) *and $f$ is a function belonging to $L_p(G)$, $p > 1$. Suppose, moreover, that the coefficients of $A$ verify condition $\{j; K\}$, $1 \leq j \leq 2m$, and that $G$ is of class $C^{2m}$. Then, $u \in H_{j,L_p}(G)$ and*

$$\text{(8.8)} \qquad \| u \|_{j, L_p(G)} \leq c \left( \| f \|_{L_p(G)} + \| u \|_{L_p(G)} \right),$$

*where $c$ is a constant depending only on $n, m, p, K, \lambda$ and the domain.*

*Proof:* From (8.7) we obtain the inequality

$$\text{(8.9)} \qquad |(u, Av)_G| \leq \| f \|_{L_p(G)} \| v \|_{L_{p'}(G)}$$

for all functions $v \in \overset{\bullet}{C}{}^{2m}(\bar{G}; \{D^\alpha\}_{|\alpha| \leq m-1})$, and the result follows by Theorem 8.1.

A case of special interest is

THEOREM 8.2′. *If the conditions of Theorem* 8.2 *hold with $j = 2m$ (i. e. if $a_\alpha \in C^{|\alpha|}(G)$ for $|\alpha| > 0$, $a_{(0, \dots, 0)}$ being bounded) then $u \in H_{2m, L_p}(G; \{D^\alpha\}_{|\alpha| \leq m-1})$ and satisfies* (8.6) *in the strong $L_p$ sense.*

*Proof*: We know already that $u \in H_{2m, L_p}(G)$ and consequently that $A^* u = f$ in the strong $L_p$ sense. To complete the proof we need only show that the trace of $D^\alpha u$ on the boundary (considered as an element of $L_1(\partial G)$) is zero for $0 \leq |\alpha| \leq m-1$. For functions $u$ which are sufficiently smooth this follows in a well known manner from (8.7). With some precautions the proof for functions of class $H_{2m, L_p}(G)$ is similar. For the sake of completeness we present a formal proof.

It suffices to show that given $x^0 \in \partial G$ there exists a neighborhood $\pi$ of $x^0$ on $\partial G$ such that the traces $\gamma(D^\alpha u)$ $(|\alpha| \leq m-1)$ are zero when restricted to $\pi$. Since the last property remains invariant under a domain homeomorphism of class $C^{2m}$ we may assume without loss of generality that $\pi$ is the $n-1$ dimensional sphere $x_n = 0$, $|x'| < r$ $(x' = (x_1, \ldots, x_{n-1}))$. We may also assume that the cylinder: $x' \in \pi$, $0 < x_n < \delta$, for some $\delta > 0$, belongs to $G$. Furthermore, noting that the trace $\gamma(D_n^j u) \in H_{2m-1-j, L_p}$ on $\pi$ (this follows from the estimate (2.3)), and that

$$\gamma(D_{x'}^{\alpha} D_n^j u) = D_{x'}^{\alpha} \gamma(D_n^j u) \quad \text{on} \quad \pi \tag{8.10}$$

for all derivatives $D_{x'}^{\alpha}$ of order $|\alpha| \leq 2m-1-j$ not involving $x_n$, we conclude that it will suffice to show that $\gamma(D_n^j u)$ is a null-function on $\pi$ $(j = 0, \ldots, m-1)$.

Let $0 \leq j \leq m-1$ and assume that $\gamma(D_n^k u)$ is a nullfunction on $\pi$ for every $k \leq j-1$ (there is no assumption for $j = 0$). We shall show that $\gamma(D_n^j u)$ is also a nullfunction on $\pi$ and the result will follow by induction. Let $\varphi(x') \in C_0^\infty(\pi)$ and let $\zeta(x_n)$ be a $C^\infty$ function on $x_n \geq 0$ such that $\zeta \equiv 0$ for $x_n \geq \delta$, $\zeta(x_n) = (-x_n)^{2m-1-j}/(2m-1-j)!$ for $0 \leq x_n \leq \delta/2$. Put:

$$w(x', x_n) = \varphi(x')\, \zeta(x_n).$$

Since $u \in H_{2m, L_p}(G)$ we can integrate $(u, A w)_G$ by parts to obtain the usual Green's formula with boundary values taken in the generalized trace sense. A simple calculation shows that

$$(u, A w)_G = (A^* u, w)_G + \int_\pi \gamma(D_n^j(a u))\, \varphi(x')\, dx' \tag{8.11}$$

where a is the coefficient of $D_n^{2m}$ in $A$. Since $A^* u = f$ and $(u, A w)_G = (f, w)_G$; we conclude from (8.11) that

$$\int_\pi \gamma(D_n^j(a u))\, \varphi(x')\, dx' = 0$$

for all $\varphi \in C_0^\infty(\pi)$. This implies that $\gamma(D_n^j(a u))$ and consequently $\gamma(D_n^j u)$ are null functions on $\pi$ (since $a \neq 0$), and completes the proof.

Suppose that the function $f$ in Theorem 8.2′ belongs to $L_\infty(G)$. Then, by the theorem, $u \in H_{2m, L_p}(G)$ for every $p$ so that, using Sobolev's inequalities, $u \in C^{2m-1}(\overline{G}; \{D^\alpha\}_{|\alpha|\le m-1})$ [10]. If, moreover, $f \in C^1(G)$ and $a_\alpha \in C^{|\alpha|+1}(G)$ then, by Theorem 7.1′, $u$ also belongs to $C^{2m}(G)$ [11]. Thus, in this case $u$ is an ordinary solution of (8.6).

As a side application of Theorem 8.2′ we mention

THEOREM 8.3. *Let $u$ be a function belonging to $C^{2m}(G) \cap C^{m-1}(\overline{G})$, such that:*

$$\begin{cases} Au = 0 & in \quad G, \\ D^\alpha u = 0 & on \quad \partial G, \quad |\alpha| \le m-1, \end{cases} \tag{8.12}$$

*where $A$ is the elliptic operator* (8.1). *If $G$ is of class $C^{2m}$ and the coefficients $a_\alpha \in C^{|\alpha|}(G)$, then $u \in H_{2m,L_p}(G)$ for every $p$.*

To prove Theorem 8.3 it suffices to show that

$$(u, A^*v)_G = 0$$

for all functions $v \in \dot{C}^{2m}(G; \{D^\alpha\}_{|\alpha|\le m-1})$, the result will then follow from Theorem 8.2′. This, however, follows from (8.12) by Green's formula applied to $u$ and $v$ in $G$ (using a suitable approximation procedure).

Theorem 8.3 is useful in connection with uniqueness theorems for the Dirichlet problem (for strongly elliptic equations) where it is necessary to assume in general that $u \in H_{m,L_2}(G)$. The theorem shows that this extra condition is really superfluous [12].

The main applications of the regularity theorems will be given in Part II. In conclusion we add only the following

REMARK: Combining Theorem 8.2 with the a priori $L_p$ estimates up to the boundary given in Agmon-Douglis-Nirenberg [3] one can show (with suitable regularity assuptions on the domain and the coefficients of $A$) that if in Theorem 8.2 $f \in H_{k,L_p}(G)$, then $u \in H_{2m+k,L_p}(G)$. Similarly, using the Schauder estimates in integral form of [3] one can show that if $f$ belongs to the Hölder class $C^{k+\mu}(\overline{G})$ ($k$ a non-negative integer and $0 < \mu < 1$), then $u \in C^{2m+k+\mu}(\overline{G})$.

---

(10) More precisely, $u$ belongs to the Hölder class $C^{2m-1+\mu}(\overline{G})$ for every $\mu < 1$.

(11) We note that the same conclusion holds with the following weaker assumptions: $f$ and $a_{(0,\ldots,0)}$ are continuous functions satisfying locally a Hölder condition $a_\alpha \in C^{|\alpha|}(G)$ for $|\alpha| > 0$. For a proof see Agmon-Douglis-Nirenberg [3; Appendix 5].

(12) In this connection see also Miranda [20] Lemma 11.1.

# BIBLIOGRAPHY I


[1] AGMON. S., *Multiple layer potentials and the Dirichlet problem for higher order elliptic equations in the plane I,* Comm. Pure Appl. Math., Vol. 10, 1957, pp. 179-240.

[2] AGMON, S., *The coerciveness problem for integro-differential forms, J.* d'Analyse Math., Vol. 6, 1958, pp. 183-223.

[3] AGMON, S., DOUGLIS, A. and NIRENBERG, L., *Estimates near the boundary for solutions of elliptic partial differential equations satisfying general boundary conditions I,* Comm. Pure Appl. Math., Vol. 12, No. 4, 1959.

[4] ARONSZAJN, N., *On Coercive integro-differential quadratic forms,* Conference of Partial Differential Equations, University of Kansas, Technical Report No. 14, pp. 94-106.

[5] BABICH, V. M., *On a problem of extension of functions,* Uspehi Mat. Nauk. N. S., Vol. 8, no. 2 (54), 1953, pp. 111-113 (in Russian).

[6] BROWDER F. E., *Strongly elliptic systems of differential equations,* Ann. Math. Study No. 33. Princeton University Press, 1954, pp. 15-51.

[7] BROWDER, F. E., *On the regularity properties of solutions of elliptic partial differential equations,* Comm. Pure Appl. Math., Vol. 9, 1956, pp. 351-361.

[8] CALDERON A. P. and ZIGMUND, A., *On singular integrals,* Amer. J. Math., Vol. 78, 1956, pp. 289-309.

[9] DENY J. and LIONS J. L., *Les espace de Beppo Levi,* Ann. Inst. Fourier, Vol. 5, 1953-54, pp. 305-370.

[10] FICHERA, G., *Alcuni recenti sviluppi della teoria dei problemi al contorno per le equazioni alle derivate parziali lineari,* Atti Cong. Int. Equaz. Lin. der. parz. di Trieste, 1954, pp. 174-227.

[11] FRIEDRICHS, K. O., *The identity of weak and strong extensions of differential operators,* Trans. Amer. Math. Soc., Vol. 55, No. 1, 1944, pp. 132-151.

[12] FRIEDRICHS, K. O., *On the differentiability of solutions of linear elliptic differential equations,* Comm. Pure Appl. Math., Vol. 6, 1953, pp. 299-326.

[13] GAGLIARDO, E., *Proprietà di alcune classi di funzioni in più variabili,* Ric. di Mat., Vol. 7, 1958, pp. 102-137.

[14] GÅRDING, L., *Dirichlet's problem for linear elliptic partial differential equations,* Math. Scand., Vol. 1, 1953, pp. 55-72.

[15] HÖRMANDER, L., *Differentiability properties of solutions of systems of differential equations,* Ark. for Mat. Vol. 3, 1958, pp. 527-535.

[16] JOHN, F., *Plane waves and spherical means applied to partial differential equations,* Interscience. New York, 1955.

[17] KOSELEV, A. I., *On boundedness in $L_p$ of derivatives of solutions of elliptic equations and systems,* Doklady Akad. Nauk USSR, Vol. 116, 1957, pp. 542-544 (in Russian).

[18] LIONS, J. L., *Problemès aux limites en theorie des distributions,* Acta Math., Vol. 94, 1955, pp. 13-153.

[19] MAGENES, E. and STAMPACCHIA, G., *I problemi al contorno per le equazioui differenziali di tipo ellittico,* Ann. Sc. Norm. Sup. Pisa Vol. 12, 1958, pp. 247-358.

[20] MIRANDA, C., *Teorema del massimo modulo e teorema di essistenza e di unicita per il problema di Dirichlet redativo alle equazioni. ellittiche in due variabili,* Ann. Mat. Pura Appl., Vol. 46, 1958, pp. 265-312.

[21] MORREY, C. B. *Functions of several variables and obsolute continuity,* Duke Math. J., Vol. 6, 1940, pp. 187-215.

[22] MORREY, C. B., *Second order elliptic systems of differential equations,* Contributions to the Theory of Partial Differential Equations, Ann. Math. Studies No. 33 Princeton, 1954, pp. 101-159.

[23] NIRENBERG, L., *Remarks on strongly elliptic partial differential equations,* Comm. Pure Appl. Math., Vol. 8, 1955, pp. 648-674.

[24] NIRENBERG, L., *Estimates and existence of solutions of elliptic equations,* Comm. Pure Appl. Math., Vol. 9, 1956, pp. 509-530.

[25] SCHECHTER, M., *Integral inequalities for partial differential operators and functions satisfying general boundary conditions,* Comm. Pure Appl. Math., Vol. 12, 1959, pp. 37-66.

[26] SCHECHTER, M., *Solution of the Dirichlet problem for systems not necessarily strongly elliptic,* Comm. Pure Appl. Math., Vol. 12, 1959, pp. 241-247.

[27] SCHECHTER, M., *General boundary value problems for elliptic partial differential equations,* Bull. Amer Math. Soc., Vol. 65, 1959, pp. 70-72.

[28] SCHWARTZ, L , *Théorie des distributions,* Hermann, Paris, 1950-51.

[29] SCHWARTZ, L., *Su alcuni problemi della teoria delle equazioni differenziali lineari di tipo ellittico,* Rend. Sem. Mat. Fis. Milano, Vol. 27, 1958, pp. 211-249.

[30] SOBOLEV S. L., *On a theorem of functional analysis,* Mat. Sbornik, N. S. 4, 1938, pp. 471-497 (in Russian).

[31] STAMPACCHIA, G., *Sopra una classe di funzioni in n variabili,* Ric. di Mat., Vol. 1, 1951, pp. 27-54.

[32] VISHIK, M. I., *On strongly elliptic systems of differential equations,* Mat. Sbornik, Vol. 29, 1951, pp. 617-676 (in Russian).

The Hebrew University
Jerusalem, Israel

GUBBIO - SOC. TIPOGRAFICA «ODERISI» 1960

Estratto dagli *Annali della Scuola Normale Superiore di Pisa*
Serie III. Vol. XIV. Fasc. I (1960)

# MULTIPLE INTEGRAL PROBLEMS IN THE CALCULUS OF VARIATIONS AND RELATED TOPICS

by

CHARLES B. MORREY, JR. (Berkeley) (*)

## Introduction.

In this series of lectures, I shall present a greatly simplified account of some of the research concerning multiple integral problems in the calculus of variations which has been reported in detail in the papers [39], [40], [41], [42], [44], [46], and [47]. I shall speak only of problems in non-parametric form and shall therefore not describe the excellent result concerning double integrals in parametric form obtained almost concurrently by Sigalov, Danskin, and Cesari [62], [9], [5]) nor the work of L. C. Young and others on generalized surfaces. Some of my results have been extended in various ways by Cinquini [6], De Giorgi [10], Fichera [17], Nöbéling [51], Sigalov [58], [59], [60], [61], Silova [63], and Stampacchia [67], [68], [69], [70]. However, it is hoped that the results presented here will serve as an introduction to the subject.

The first part of this research reported in these lectures is an extension of Tonelli's work on single and double integral problems in which he employed the so-called direct methods of the calculus of variations ([71] through [78]). His work was stimulated, no doubt, by the succes of Hilbert, Lebesgue [31] and others in the rigorous establishment of Dirichlet's principle in certain important cases. The principle idea of these direct methods is to establish the existence of a function $z$ minimizing an integral by showing (i) that the integral $I(z)$ is lower semicontinuous with respect to some

---

(*) Presented at the international conference organized by C.I.M.E in Pisa, september 1-10-1958.

kind of convergence, (ii) that $I(z) \geq d$ for the $z$ considered and (iii) that there is a «minimizing sequence» $z_n$ such that $I(z_n) \to d$ and $z_n \to z_0$ in the sense required.

In the case of single integral problems, where

$$(0.1) \qquad I(z) = \int_a^b f[x, z(x), z'(x)]\, dx$$

Tonelli (see, for instance [76]) was ableto carry through this program for the case that only absolutely continuous functions are admitted, the convergence is uniform, and (essentially) $f(x, z, p)$ is convex in $p$ (if $f(x, z, p) \geq f_0(p)$ where $f_0(p)/|p| \to \omega$, it is seen from the proof of Theorem 2.4 below, that the functions in any minimizing sequence would be uniformly absolutely continuous so that a subsequence would converge uniformly to an absolutely continuous function $z_0$ which would thus minimize $I(z)$). Tonelli was also able to carry through the entire program for certain double integral problems using functions absolutely continuous in his sense (ACT) and uniform convergence [77], [78]. However, in general he had to assume that the integrand $f(x, y, z, p, q)$ satisfied a condition like

$$(0.2) \qquad f(x, y, z, p, q) \geq m(p^2 + q^2)^{\alpha/2} - k, \{\alpha > 2, m > 0\}.$$

If $f$ satisfies this condition, Tonelli showed that the functions in any minimizing sequence are equicontinuous, and uniformly bounded on interior domains at least (see Lemma 4.1) and so a subsequence converges uniformly on such domains to a function still in his class. He was also able to handle the case where

$$(0.3) \qquad f(x, y, z, p, q) \geq m(p^2 + q^2) - k \text{ if } f(x, y, z, 0, 0) \equiv 0,$$

for instance by showing that any minimizing sequence can be replaced by one in which each $z_n$ is monotone in the sense of Lebesgue (see [31] and [37], for instance) and hence equicontinuous on interior domains, etc.

However, Tonelli was not able to get a general theorem to cover the case where $f$ satisfies (0.2) only with $1 < \alpha < 2$. Moreover, if one considers problems involving $\nu > 2$ independent variables, one soon finds that one would have to require $\alpha$ to be $> \nu$ in (0.2) in order to ensure that the functions in any minimizing sequence would be equicontinuous on interior domains. To see this, one needs only to notice that the functions

$$\log\log(1 + 1/r), \quad 1/r^h, 0 < r \leq 1 \; (r^2 = \sum_{\alpha=1}^{\nu} (x^\alpha)^2),$$

are limits of ACT functions in which

$$\int_{B(0,1)} | \nabla z_n |^{\nu} dx \text{ and} \int_{B(0,1)} | \nabla z_n |^{k} dx \text{ for } k < \nu/(h+1)$$

respectively, are uniformly bounded (see below for notation).

In order to carry through the program, for these more general problems, then, the writer found it expedient to allow functions which are still more general than Tonelli's ACT functions. One obtains these more general functions by merely replacing the requirement of $\nu$-dimensional continuity in Tonelli's definition by summability, but retaining Tonelli's requirements of absolute continuity along lines parallel to the axes, summable partial derivatives, etc. But then, two such functions may differ on a set of measure zero in such a way that their partial derivatives also differ only on a set of measure zero. It is clear that such functions should be identified and this in done in forming the « spaces $\mathcal{B}_\lambda$ » discussed in Chapter I.

These more general functions have been defined in various ways and studied by various authors in various connections. Beppo Levi [32] was probably the first to use functions of this type in the special case that the function and its first derivatives are in $\mathcal{L}_2$; any function equivalent to such a function has been called strongly differentiable by Friedrichs and these functions and those of corresponding type involving higher derivatives have been used extensively in the study of partial differential equations (see [2], [3], [11], [18], [19], [20], [21], [24], [28], [30], [42], [45], [46], [47], [50], [57], [61], [66]), G. C. Evans also made use at an early date [14], [15], [16] of essentially these same functions in connection with his work on potential theory. J. W. Calkin needed them in order to apply Hilbert space theory to the study of boundary value problems for elliptic partial differential equations and collaborated with the author in setting down a number of useful theorems about these functions (see [4] and [40]). The functions have been studied in more detail since the war by some of the writers mentioned above and by Aronszajn and Smith who showed that any function in the space $H_{m0}$ (see Professor Niremberg's lectures) can be represented as a Riesz potential of order $m$ [1]. The writer is sure that many others have also discussed these functions and certainly does not claim that the bibliography is complete.

In Chapter I, the writer presents some of the known results concerning these more general functions. In Chapter II, these are applied to obtain theorems concerning the lower-semicontinuity and existence of minima of multiple integrals of the form

$$I(z, G) = \int_G f[x, z(x), \nabla z(x)]\, dx$$

(0.5) $$x = (x^1, \ldots, x^\nu),\ z = (z^1, \ldots, z^N),\ \nabla z = \{\partial z^i/\partial x^\alpha\},\ dx = dx^1 \ldots dx^\nu$$

$$i = 1, \ldots, N,\ \alpha = 1, \ldots, \nu$$

where the function $f$ is assumed to be continuous in $(x, z, p)$ for all $(x, z, p)$ and convex in $p = \{p^i_\alpha\}$ for each $(x, z)$. In Chapter III, the most general type of function $f(x, z, p)$ for which the integral $I(z, G)$ in (0.5) is lower-semicontinuous is discussed. In Chapter IV, the writer discusses his results concerning the differentiability of the solutions of minimum problems. In Chapter V, the writer discusses the recent application by Eells and himself of a variational method in the theory of harmonic integrals.

We consistently use the notations of (0.5). Il $\varphi$ is a vector, $|\varphi|$ denotes the square root of the sum of the squares of the components. Our functions are all real-valued unless otherwise noted. If $z$ is a vector or tensor $z_\alpha, z_{\alpha\beta}$, etc., will denote the partial derivatives $\partial z/\partial x^\alpha$, $\partial^2 z/\partial x^\alpha \partial x^\beta$, etc., or their corresponding generalized derivatives. Repeated indices are summed unless otherwise noted. If $G$ is a domain $\partial G$ denotes its boundary and $\bar{G} = G \cup \partial G$. $B(x_0, R)$ denotes the solid sphere with center at $x_0$ and radius $R$; we sometimes abbreviate $B(0, R)$ to $B_R$, $[a, b]$ denotes the closed cell $a^\alpha \leq x^\alpha \leq b^\alpha$. All integral are Lebesgue integrals. It is sometimes desirable to consider the behavior of a function (or vector) $z(x)$ with respect to a particular variable $x^\alpha$; when this is done, we write $x = (x^\alpha, x'_\alpha)$ and $z(x) = z(x^\alpha, x'_\alpha)$ where $x'_\alpha$ stands the remaining variables; sometimes $(\nu - 1)$ dimensional integrals

$$\int_{a'_\alpha}^{b'_\alpha} f(x^\alpha_0, x'_\alpha)\, dx'_\alpha$$

appear in which case they have their obvious significance. We say that a (vector) function $z(x)$ satisfies a uniform Lipschitz condition on a set $S$ if and only if there is a constant $M$ such that

$$|z(x_1) - z(x_2)| \leq M \cdot |x_1 - x_2| \text{ for } x_1 \text{ and } x_2 \text{ on } S;$$

$z$ is said to satisfy a uniform Hölder condition on $S$ with exponent $\mu$, $0 < \mu < 1$, if and only if there is an $M$ such that

$$|z(x_1) - z(x_2)| \leq M \cdot |x_1 - x_2|^\mu \text{ for } x_1 \text{ and } x_2 \text{ on } S.$$

A (vector) function $z$ is of class $C^n$ on a domain $G$ if and only if $z$ and its partial derivatives of order $\leq n$ are continuous on $G$; $z$ is said to be of class $C^{n+\mu}$ or $C_\mu^n$ on $G$ if and only if $z$ is of class $C^n$ on $G$ and it and all of its partial derivatives of order $\leq n$ satisfy uniform Hölder conditions with exponent $\mu$, $0 < \mu < 1$, on $G$; the second notation $C_\mu^n$ is used when $\mu = 1$ (see Chapter V).

## CHAPTER I

### Function of class $\mathcal{B}_\lambda$, $\mathcal{B}'_\lambda$, $\mathcal{B}''_\lambda$ $(\lambda \geq 1)$ and functions which are ACT.

We begin with the definitions of these classes:

DEFINITION: A function $z(x)$ $(x = (x^1, \dots, x^\nu))$ *is of class* $\mathcal{B}_\lambda$ *on a domain* $G$ if and only if $z$ is of class $\mathcal{L}_\lambda$ on $G$ and there are functions $p_\alpha$, $\alpha = 1, \dots, \nu$, of class $\mathcal{L}_\lambda$ on $G$ with the following property; if $R$ is any cell with closure in $G$, there is a sequence $z_{nR}$ of functions of class $C'$ on $R \cup \partial R$ such that $z_n \to z$ and $z_{n,\alpha} \to p_\alpha$ strongly in $\mathcal{L}_\lambda$ on $R$.

DEFINITION: A function $z$ *is of class* $\mathcal{B}'_\lambda$ on $G$ if and only if

(i) $z$ is of class $\mathcal{L}_\lambda$ on $G$;

(ii) if $[a, b]$ is any closed cell in $G$, then $z$ is AC (absolutely continuous) in $x^\alpha$ on $[a^\alpha, b^\alpha]$ for almost all $x'_\alpha$ on $[a'_\alpha, b'_\alpha]$, $\alpha = 1, \dots, \nu$;

(iii) the partial derivatives $z_{,\alpha}$, which exist almost every-where and are measurable on account of (ii), are of class $\mathcal{L}_\lambda$ on $G$.

DEFINITION: A function $z$ *is of class* $\mathcal{B}''_\lambda$ on $G$ if and only if $z$ is of class $\mathcal{B}_\lambda$ on $G$ and is continuous there.

DEFINITION: A function $z$ is *absolutely continuous in the sense of Tonelli* (ACT) on $G$ if and only if $z$ is of class $\mathcal{B}'_1$ and is continuous on $G$.

DEFINITION: Suppose $z$ is of class $\mathcal{L}_1$ on $G$. We define its $h$ average function on the set $G_h$ by

$$z_h(x) = (2h)^{-\nu} \int_{x-h}^{x+h} z(\xi)\, d\xi, \tag{1.1}$$

$G_h$ being the set of all $x$ in $G$ such that the cell $[x - h, x + h] \subset G$.

LEMMA 1.1: *Is $z$ is of class $\mathcal{L}_\lambda$ on a domain $G$ and $z_h$ is its $h$-average function defined on $G_h$, then $z_h \to z$ in $\mathcal{L}_\lambda$ as $h \to 0$ on each closed cell $[a, b]$ in $G$ and $z_h$ is continuous on $G_h$.*

*Proof*: That $z_h$ is continuous follows from the absolute continuity of the Lebesgue integral. Next, it is well known that $z_h(x) \to z(x)$ as $h \to 0$ for almost all $x$. Finally, choose $h_0 > 0$ so that $[a - h_0, b + h_0] \subset G$, keep $0 < h < h_0$, and let $\varphi(\varrho)$ be a function $\to 0$ as $\varrho \to 0$ such that $\| z \|_e \leq \varphi[m(e)]$ for $e \subset [a - h_0, b + h_0]$, where

$$\| z \|_e = \left[ \int_e | z(x) |^\lambda \, dx \right]^{1/\lambda}.$$

Then the lemma follows, since

$$\| z_h - z \|_e \leq \| z_h \|_e + \| z \|_e \leq 2\varphi\,[m\,(e)] \text{ for } e \subset [a\,,b]$$

since

$$\int_e | z_h\,(x) |^\lambda\, dx \leq (2h)^{-\nu} \int_{-h}^{h} \left[ \int_{e(\xi)} | z\,(x + \xi) |^\lambda\, dx \right] d\xi =$$

$$= (2h)^{-\nu} \int_{-h}^{h} \left[ \int_{e(\xi)} | z\,(y) |^\lambda\, dy \right] d\xi \leq \left\{ \varphi\,[m\,(e)] \right\}^\lambda .$$

where $e\,(\xi)$ is the set obtained by translating $e$ along the vector $\xi$.

THEOREM 1.1: *If $z$ is of class $\mathcal{B}_\lambda$ on $G$, the functions $p_\alpha$ are uniquely determined up to null functions. If $z_h$ is the h average of $z$ and $p_{\alpha h}$ is that of $p_\alpha$, then $z_h$ is of class $C'$ on $G_h$* and

$$z_{h,\alpha}\,(x) = p_{\alpha h}\,(x)\,, \qquad h > 0. \tag{1.2}$$

*Proof*: Let $[a\,,b] \subset G$, choose $h_0$ so $[a - h_0\,, b + h_0] \subset G$, and keep $0 < h < h_0$. Approximate to $z$ and $p_\alpha$ by $z_n$ and $z_{n,\alpha}$ in $\mathcal{L}_\lambda$ on $[a - h_0\,, b + h_0]$. Then for each $h$, we see that $z_{nh;\alpha} = (z_{n,\alpha})_h$ and we may obtain (1.2) by letting $n \to \infty$ on $[a\,,b]$. The first statement is now obvious.

DEFINITION: If $z$ is of class $\mathcal{B}_\lambda$ on a domain $G$, we define its *generalizet derivative* $D_\alpha\, z\,(x)$ as the Lebesgue derivative at $x$ of the set function

$$\int_e p_\alpha\,(x)\, dx\,.$$

THEOREM 1.2: *If $z$ is of class $\mathcal{B}'_\lambda$ on $G$, $z_h$ is its h-average function, and $p_{\alpha h}$ is that of its partial derivative $\partial z/\partial x^\alpha$, then $z_h$ is of class $C'$ and* (1.2) *holds. Moreover $z$ is of class $\mathcal{B}_\lambda$ and its corresponding partial and generalized derivatives coincide almost everywhere.*

*Proof*: Let $[a\,,b] \subset G$, choose $h_0$ so $[a - h_0\,, b + h_0] \subset G$, and keep $0 < h < h_0$. If $x'_\alpha$ is not in *a* set of measure 0 on $[a'_\alpha - h_0\,, b'_\alpha + h]$, then $\partial z/\partial x^\alpha \equiv p_\alpha$ is summable in $x^\alpha$ over $[a^\alpha - h_0\,, b^\alpha + h_0]$ and

$$\int_{x_1^\alpha}^{x_2^\alpha} p_\alpha\,(x^\alpha\,, x'_\alpha)\, dx^\alpha = z\,(x_2^\alpha\,, x'_\alpha) - z\,(x_1^\alpha\,, x'_\alpha)\,. \tag{1.3}$$

By integrating (1.3), we see that it holds for all $x'_\alpha$ on $[a'_\alpha\,, b'_\alpha]$ and all $x_1^\alpha\,, x_2^\alpha$ on $[a^\alpha\,, b^\alpha]$ if $z$ and $p_\alpha$ are replaced by their $h$-averages. Then (1.2) and the last statement follow.

THEOREM 1.3: (a) *If $z_1$ and $z_2$ are equivalent and one is of class $\mathfrak{B}^\lambda$ on $G$, then both are and their generalized derivatives coincide.*

(b) *If $z_1$ and $z_2$ are of class $\mathfrak{B}_\lambda$ on a domain $G$ and $z_{1,\alpha}(x) = z_{2,\alpha}(x)$ almost everywhere on $G$, then $z_1$ and $z_2$ differ by a constant and a null function.*

These are easily proved using the $h$-average functions.

THEOREM 1.4: (a) *Any function $z$ of class $\mathfrak{B}_\lambda$ on $G$ is equivalent to a function $z_0$ of class $\mathfrak{B}'_\lambda$ on $G$.*

b) *$z$ is ACT on $G$ if and only if $z$ is of class $\mathfrak{B}''_1$ there.*

*Proof*: To prove (a), let $R = [a, b]$ be any rational cell in $G$ and approximate to $z$ there by functions $z_n$ of class $C'$ on $[a, b]$. A subsequence, still called $z_n$, converges to $z$ almost everywhere and is such that

$$(1.4) \qquad \lim_{n\to\infty} \int_{a^\alpha}^{b^\alpha} | z_{n,\alpha}(x^\alpha, x'_\alpha) - z_{,\alpha}(x^\alpha, x'_\alpha) |^\lambda \, dx^\alpha = 0$$

for all $x'_\alpha$ not in a set $Z_{R\alpha}$ of $(\nu - 1)$-dimensional measure zero, $\alpha = 1, \ldots, \nu$. From (1.4), we see that the $z_n(x^\alpha, x'_\alpha)$ are equicontinuos in $x^\alpha$ and converge uniformly on $[a^\alpha, b^\alpha]$ to a function $z_{0R}(x^\alpha, x'_\alpha)$ which is $AC$ in $x^\alpha$ if $x'_\alpha$ is not in $Z_{R\alpha}$, $\alpha = 1, \ldots, \nu$. Obviously $z_{0R} = z$ almost everywhere on $R$. Since the union of the $Z_{R\alpha}$ for $\alpha$ fixed and $R$ running over all rational cells is still of measure zero; we see that the $z_{0R}$ join up to form a function $z_0$ of class $\mathfrak{B}'_\lambda$ on $G$.

To prove (b), we note first that if $z$ is $ACT$ on $G$, it is of class $\mathfrak{B}''_1$ on $G$. Conversely, if $z$ is of class $\mathfrak{B}''_1$, we may repeat the first part of the proof taking $z_n$ as the $h_n$-average of $z$ and conclude that we may take $z_{0R}$ always $= z$ since then $z_n$ converges uniformly to $z$ on $R$.

The following theorems are easily proved by approximations:

THEOREM 1.5: *The space $\mathfrak{B}_\lambda$ of equivalence classes of functions of class $\mathfrak{B}_\lambda$ is a Banach space if we define the norm by*

$$\| z \|_\lambda = \left\{ \int_G \left[ |z|^2 + \sum_{\alpha=1}^{\nu} |z_{,\alpha}|^2 \right]^{\lambda/2} dx \right\}^{1/\lambda}.$$

*If $\lambda = 2$, $\mathfrak{B}_\lambda$ is a real Hilbert space if we define*

$$(z, w) = \int_G \left( zw + \sum_{\alpha=1}^{\nu} z_{,\alpha} w_{,\alpha} \right) dx.$$

THEOREM 1.6: *If $u \in \mathcal{B}_\lambda$ and $h$ is of class $C'$ and satisfies a uniform Lipschitz condition on the bounded domain $G$, then $hu \in \mathcal{B}_\lambda$ on $G$ and the generalized derivatives $(hu)_{,\alpha}$ all exist at any point $x_0$ where all the $u_{,\alpha}(x_0)$ exist.*

DEFINITION: A transformation $T: x = x(y)$ from a domain $\widetilde{G}$ onto $G$ which is of class $C'$ is said to be *regular* if and only if $T$ is $1-1$ and $T$ and its inverse are of class $C'$ and satisfy uniform Lipschitz condition ($|x(y_1) - x(y_2)| \leq M \cdot |y_1 - y_2|$, etc.).

THEOREM 1.7: *If $u$ is of class $\mathcal{B}_\lambda$ $(\mathcal{B}_\lambda'')$ on the bounded demain $G$, $x=x(y)$ is a regular transformation of class $C'$ from the bounded domain $\widetilde{G}$ onto $G$ and $\widetilde{u}(y) = u[x(y)]$, then $\widetilde{u}$ is of class $\mathcal{B}_\lambda$ $(\mathcal{B}_\lambda'')$ on $\widetilde{G}$. Moreover, if $x_0 = x(y_0)$ and all the generalized derivatives $u_{,\alpha}(x_0)$ exist, then all the generalized derivatives $\widetilde{u}_{,\alpha}(y_0)$ exist and*

$$\widetilde{u}_{,\beta}(y_0) = u_{,\alpha}[x(y_0)] \cdot x^\alpha_{,\beta}(y_0) \tag{1.5}$$

*Proof*: That $\widetilde{u}$ is of class $\mathcal{B}_\lambda$ $(\mathcal{B}_\lambda'')$ and that we may choose the right sides of (1.5) as the «derivative functions» $\widetilde{p}_\beta$ of the definition is easily proved by approximating $u$ on interior domains by functions of class $C'$. Since regular families of sets correspond under regular transformations, the last statement follows easily.

REMARKS: It is proved in [40] and [47], for instance, that if $u$ is of class $\mathcal{B}_\lambda$ on $G$, it is equivalent to a function $\overline{u}$ (namely the Lebesgue derivetive of $\int_e u\, dx$) which is of class $\mathcal{B}_\lambda'$ and is such that any transform as in Theorem 1.7 retains this property. But the last statement of Theorems 1.7 does not hold for the partial derivatives since this would imply that $z$ had a total differential almost everywhere contrary to an example of Sake [55]. It is clear how to define the generalized derivative in a given direction and that (Theorem 1.7) if all the $u_{,\alpha}(x_0)$ exist, then a all the *generalized* directional derivatives exist at $x_0$ and are given by their usual formulas there. It is now easy to prove Rademacher's famous theorem [52] that a Lipschitz function has a total differential almost everywhere: For using the result just mentioned together with Theorem 1.2 we see that if $z$ is Lipschitz and $x_0$ is not in a set of measure zero, then the *partial* and generalized derivatives all exist at $x_0$ and the ordinary directional derivatives in a denumerable everywhere dense set of directions (independent of $x_0$) all exist and are given by their usual formulas; at any such point $z$ is seen to have a total differential. Thus in Theorem 1.6, $h$ may be Lipschitz and in Theorem 1.7, the transformation and its inverse may be Lipschitz; in this case (1.5) holds whenever all the *generalized* derivatives involved exist.

THEOREM 1.8: *The most general linear functional on the space* $\mathfrak{B}_\lambda$ *is of the form*

$$f(x) = \int_G \left(A_0 z + \sum_{\alpha=1}^{\nu} A_\alpha z_{,\alpha}\right) dx \tag{1.6}$$

*where the* $A_\alpha$ $(\alpha \geq 0) \in \mathfrak{L}_\mu$ *with* $\lambda^{-1} + \mu^{-1} = 1$ *if* $\lambda > 1$ *or are bounded and measurable on* $G$ *if* $\lambda = 1$.

*Proof*: Let $A_\lambda$ be the space of all vectors $\varphi = (\varphi_0, \ldots, \varphi_\nu)$ with components in $\mathfrak{L}_\lambda$ and

$$\| \varphi \| = \left\{ \int_G \left[ \sum_{\alpha=0}^{\nu} \varphi_\alpha^2 \right]^{\lambda/2} dx \right\}^{1/\lambda} .$$

From Theorem 1.5 it follows that the subspace of all vectors $(z, z_{,1}, \ldots, z_{,\nu})$ for which $z \in \mathfrak{B}_\lambda$ on $G$ is a closed linear manifold $M$ in $B_\lambda$. Hence if $F(z, z_{,1}, \ldots, z_{,\nu}) = f(z)$, then $F$ can be extended to the whole space $B$ to have same norm as $f$. Then $F$ is given by (1.6).

From Theorem 1.8 we immediately obtain:

THEOREM 1.9: (a) *A necessary and sufficient condition that* $z_n$ *converges weakly to* $z$ $(z_n \rightharpoonup z)$ *in* $\mathfrak{B}_\lambda$ *on* $G$ *is that* $z_n \rightharpoonup z$ *and the* $z_{n,\alpha} \rightharpoonup z_{,\alpha}$ *in* $\mathfrak{L}_\lambda$ *on* $G$.

(b) *If* $z_n \rightharpoonup z$ *in* $\mathfrak{B}_\lambda$ *on* $G$, *then* $z_n \rightharpoonup z$ *in* $\mathfrak{B}_\lambda$ *on any subdomain.*

(c) *If* $z_n \rightharpoonup z$ *in* $\mathfrak{B}_\lambda$ *on* $G$ (bounded), $x = x(y)$ *is a regular transformation of class* $C'$ *from* $\widetilde{G}$ *onto* $G$, $\widetilde{z}_n(y) = z_n[x(y)]$ *and* $\widetilde{z}(y) = z[x(y)]$, *then* $\widetilde{z}_n \rightharpoonup \widetilde{z}$ *in* $\mathfrak{B}_\lambda$ *on* $\widetilde{G}$.

(d) *If* $z_n \rightharpoonup z$ *in* $\mathfrak{B}_\lambda$ *on* $G$ (bounded) *and* $h$ *is Lipschitz on* $G$, *then* $hz_n \rightharpoonup hz$ *in* $\mathfrak{B}_\lambda$ *on* $G$.

DEFINITION: A function $z$ is of class $\mathfrak{B}_{\lambda 0}$ on $G$ (bounded) if and only if it is of class $\mathfrak{B}_\lambda$ there and there existe a sequence $\{u_n\}$, each of class $C'$ and vanishing on and near the boundary $\partial G$ such that $z_n \to z$ (strong convergence) in $\mathfrak{B}_\lambda$ on $G$. The *subspace* $\mathfrak{B}_{\lambda 0}$ of $\mathfrak{B}_\lambda$ is defined correspondingly. If $z$ and $z^* \in \mathfrak{B}_\lambda$ on $G$, we say thath $z = z^*$ *on* $\partial G$ *in the* $\mathfrak{B}_\lambda$ *sense* if and only if $z - z^* \in \mathfrak{B}_{\lambda 0}$ on $G$·

The follawing is immediate:

THEOREM 1.10: *The subspace* $\mathfrak{B}_{\lambda 0}$ *is a closed linear manifold is* $\mathfrak{B}_\lambda$; *if* $z_n \rightharpoonup z$ *in* $\mathfrak{B}_\lambda$ *on* $G$ *and each* $z_n \in \mathfrak{B}_{\lambda 0}$, *then* $z \in \mathfrak{B}_{\lambda 0}$. *If* $z \in \mathfrak{B}_{\lambda 0}$ *and* $z_1(x) = z(x)$ *for* $x$ *on* $G$ *and* $z_1(x) = 0$ *otherwise, then* $z_1 \in \mathfrak{B}_{\lambda 0}$ *on any* $D \supset G$ *and* $z_{1,\alpha}(x) = 0$ *for almost all* $x$ *not in* $G$.

THEOREM 1.11 (Poincaré's inequality): *Suppose* $z \in \mathfrak{B}_{\lambda 0}$ *on* $G \subset B(x_0, R)$. Then

$$\int_G |z|^\lambda dx \leq \lambda^{-1} R^\lambda \int_G |\nabla z|^\lambda dx .$$

*Proof*: It is sufficicient to prove this for $z$ of class $C'$ and vanishing on $\partial B(x_0, R)$ with $G = B(x_0, R)$. Taking spherical coordinates $(r, p)$ with $r = |x - x_0|$ and $p \in \Sigma = \partial B(0, 1)$, we obtain

$$\int_\Sigma |u(r, p)|^\lambda \, d\Sigma(p) = \int_\Sigma |u(R, p) - u(r, p)|^\lambda \, d\Sigma(p) =$$

$$= \int_\Sigma \left| \int_r^R u_r(s, p) \, ds \right|^\lambda d\Sigma \leq (R - r)^{\lambda - 1} \int_r^R \int_\Sigma |u_r(s, p)|^\lambda \, ds \, d\Sigma$$

where $u(r, p) = z(x)$. Thus

$$\int_{B(x_0, R)} |z(x)|^\lambda \, dx = \int_0^R r^{\nu - 1} \left[ \int_\Sigma |u(r, p)|^\lambda \, d\Sigma(p) \right] dr$$

$$\leq \int_0^R (R - r)^{\lambda - 1} \left[ \int_r^R s^{\nu - 1} \left\{ \int_\Sigma |u_r(s, p)|^\lambda \, d\Sigma(p) \right\} ds \right] dr$$

from which the result follows.

THEOREM 1.12: *Suppose $z \in \mathfrak{B}_\lambda$ on $G$, $\Delta \subset G$, $z^* \in \mathfrak{B}_\lambda$ on $\Delta$ and coincides with $z$ on $\partial \Delta$ in the $\mathfrak{B}_\lambda$ sense. Then the function $Z$ such that $Z(x) = z^*(x)$ on $\Delta$ and $Z(x) = z(x)$ on $G - \Delta$ is of class $\mathfrak{B}_\lambda$ on $G$ and $z_{,\alpha}(x) = z^*_{,\alpha}(x)$ almost everywhere on $\Delta$ and $Z_{,\alpha}(x) = z_{,\alpha}(x)$ almost averywhere on $G - \Delta$.*

*Proof*: For define $Z_1(x) = z^*(x) - z(x)$ on $\Delta$ and $0$ elsewhere. Then $Z(x) = z(x) + Z_1(x)$ on $G$ and the result follows from Theorem 1.10.

LEMMA 1.2: *Suppose $z \in \mathfrak{B}_\lambda$ on the cell $[a - h_0, b + h_0]$. Then*

$$\int_a^b |z_h(x) - z(x)|^\lambda \, dx \leq C_1(\nu, \lambda) \cdot h^\lambda \cdot \int_{a-h}^{b+h} |\nabla z(y)|^\lambda \, dy,$$

$$0 < h < h_0$$

*where $C_1$ depends only on the arguments indicated.*

*Proof*: Since we may approximate to $z$ strongly in $\mathfrak{B}_\lambda$ on $[a - h, b + h]$ by functions of class $C'$ on that closed cell, it is sufficient to prove the lemma for such functions. Then if $x \in [a, b]$ and $|\xi^\alpha| \leq h$, we see that $x$ and

$x+\xi$ are in $[a-h, h+b]$ so that

$$|z(x+\xi)-z(x)|^\lambda = \left|\int_0^1 \xi^\alpha z_{,\alpha}(z+t\xi)\,dt\right|^\lambda \leq |\xi|^\lambda \int_0^1 |\nabla z(x+t\xi)|^\lambda\,dt.$$

Then

$$\int_a^b |z_h(x)-z(x)|^\lambda\,dx = \int_a^b \left|(2h)^{-\nu}\int_{-h}^h [z(x+\xi)-z(x)]\,d\xi\right|^\lambda dx$$

$$\leq (2h)^{-\nu}\int_a^b\left[\int_{-h}^h |\xi|^\lambda\left\{\int_0^1 |\nabla z(x+t\xi)|^\lambda\,dt\right\}d\xi\right]dx$$

$$= (2h)^{-\nu}\int_{-h}^h |\xi|^\lambda\left\{\int_0^1\left[\int_{a+t\xi}^{b+t\xi} |\nabla z(y)|^\lambda\,dy\right]dt\right\}d\xi$$

from which the result follows.

THEOREM 1.13: *If $z_n \rightharpoonup z_0$ in $\mathfrak{B}_{0\lambda}$ on the bounded domain $G$, then $z_n \to z_0$ in $\mathfrak{L}_\lambda$ on $G$, $\lambda \geq 1$. If $\{z_n\}$ is a sequence in $\mathfrak{B}_{0\lambda}$ with $\| z_n \|$ uniformly bounded, a subsequence converges strongly in $\mathfrak{L}_\lambda$ to some function $z$.*

*Proof:* The first statement follows from the second. For, let $\{z_p\}$ be any subsequence of $\{z_n\}$. A subsequence $\{z_q\}$ converges strongly in $\mathfrak{L}_\lambda$ to some function $z$ which must be (equivalent to) $z_0$. Hence the whole sequence $z_n \to z_0$ in $\mathfrak{L}_\lambda$.

To prove the second statement, suppose $G \subset [a, b]$ and extend each $z_n$ to be 0 outside $G$; then each $z_n \in \mathfrak{B}_{0\lambda}$ on $[a-1, a+1]$ with uniformly bounded $\mathfrak{B}_\lambda$ norm. For each $h$ with $0 < h < 1$, we see that the $z_{nh}$ are uniformly bounded and equicontinuous on $[a, b]$. So there is a subsequence, called $\{z_p\}$, such that $z_{ph}$ converges uniformly to some function $z_h$ for each $h$ of a sequence $\to 0$. From lemma 1.2, it is easy to see first that the limiting $z_h$ form a Cauchy sequence in $\mathfrak{L}_\lambda$ having some limit $z$ and then that $z_p \to z$ strongly in $\mathfrak{L}_\lambda$.

In order to treat variational problems with fixed boundary values, one can, of course, practically always reduce the problem to one where the given boundary values are zero. Although one can formulate theorems about variational problems having variable boundary values on the boundary of an arbitrary bounded domain (see Chapter II), such problems become more meaningful if we restrict ourselves to domains $G$ which are bounded and of class $C'$ where boundary values can be defined in a more definite way as we now do.

DEFINITION: A bounded domain $G$ is *of class* $C'$ if and only if each point $x_0$ of the boundary $\partial G$ is interior to a neighborhood $N(x_0)$ on $G \cup \partial G$ which is the image, under a regular transformation $x = x(y)$ of class $C'$, of the half-cube $Q^+ : |x^\alpha| < 1$ for $\alpha < \nu$ and $0 \leq x^\nu < 1$, where $x(0) = x_0$ and $\partial G \cap N(x_0)$ is the image of the part of $Q^+$ where $x^\nu = 0$. Such a neighborhood $N(x_0)$ is called a ***boundary neighborhood.***

DEFINITION: Suppose $G$ is a domain. A finite sequence $\{h_1, \dots, h_N\}$ of functions is said to be a *partition of unity* of class $C'$ on $G \cup \partial G$ if and only if each $h_i$ is of class $C'$ on $G \cup \partial G$, $0 \leq h_i(x) \leq 1$ on $G \cup \partial G$ for each $i$, and

$$\sum_{i=1}^{N} h_i(x) \equiv i \quad \text{for } x \text{ on } G \cup \partial G .$$

The support of $h_i$ is the closure of the set of all $x$ on $G \cup \partial G$ for which $h_i(x) > 0$.

LEMMA 1.3: *If $G$ is bounded domain of class $C'$, there is a pattition of unity $\{h_1, \dots, h_N\}$ of class $C'$ on $G \cup \partial G$ such that the support of each $h_i$ is either interior to a cell in $G$ or is interior to a boundary neighborhood of $G \cup \partial G$.*

*Proof.* With each interior point $P$ of $G$ we define $R_P$ as the largest hypercube $|x^\alpha - x_P^\alpha| < h_P$ in $G$ and define $r_P$ as the hypercube $|x^\alpha - x_P^\alpha| < h_P/2$. With each $P$ on $\partial G$, associate a boundary neighborhood $R_P = N(P)$ which is the image under $\tau_P$ of $Q^+$ as in the definition; we define $r_P$ as the part of $R_P$ corresponding under $\tau_P$ to the part of $Q^+$ for which $|x^\alpha| < 1/2$, $\alpha = 1, \dots, \nu$. There are a finite number $r_1, \dots, r_N$ of the $r_P$ which cover $G \cup \partial G$. Clearly each corresponding $R_i$ is the image under a regular transformation $\tau_i$ of class $C'$ of either the unit cube $Q$ or the half-cube $Q^+$ where $r_i$ corresponds under $\tau_i$ to the part where $|x^\alpha| < 1/2$.

Now, let $\varphi(s)$ be a fixed function of class $C^\infty$ for all $s$ with $\varphi(s) = 1$ for $|s| \leq 1/2$, $\varphi(s) = 0$ for $|s| \geq 3/4$, and $0 \leq \varphi(s) \leq 1$ otherwise. For each $i$, define $k_i(x)$ on $R_i$ as the image under $\tau_i$ of the function $\varphi(y^1) \dots \varphi(y^\nu)$ and define $k_i(x) = 0$ elsewhere on $G \cup \partial G$. Then the support of $k_i$ is interior to $R_i$, $k_i(x) = 1$ for $x$ on $r_i$, and each $k_i$ is of class $C'$ on $G \cup \partial G$. We then define

$$h_1(x) = k_1(x), \quad h_{i+1}(x) = k_{i+1}(x) \prod_{j=1}^{i} [1 - k_j(x)], \quad i = 1, \dots, N-1 .$$

Then we see by induction that

$$h_1(x) + \dots + h_i(x) = 1 - \pi_{j=1}^{i} [1 - k_j(x)]$$

so that the sequence $\{h_i, \dots, h_N\}$ satisfies the desired conditions.

THEOREM 1.14: *Suppose $G$ is bounded and of class $C'$ and $z \in \mathfrak{B}_\lambda$ on $G$. Then*

(i) *there is a sequence $\{z_n\}$ of functions of class $C'$ on $G \cup \partial G$ which converges strongly in $\mathfrak{B}_\lambda$ to $z$ on $G$;*

(ii) *there is a boundary value function $\varphi$ in $\mathfrak{L}_\lambda$ on $\partial G$ (with respect to hyperarea) to which every sequance $\{z_n\}$ in* (i) *converges strongly in $\mathfrak{L}_\lambda$ on $\partial G$;*

(iii) *if $T: x = x(y)$ is a regular transformation of class $C'$ of $\tilde{G} \cup \partial \tilde{G}$ onto $G \cup \partial G$, $\tilde{z}(y) = z[x(y)]$, and $\tilde{\varphi}(y) = \varphi[x(y)]$, then $\tilde{\varphi}$ is the boundary value function for $\tilde{z}$ on $\partial \tilde{G}$;*

(iiii) *if $\varphi(x) = 0$ for almost all $x$ on $\partial G$, then $z \in \mathfrak{B}_{\lambda 0}$ on $G$.*

*Proof:* Let $\{h_1, \dots, h_N\}$ be a partition of unity on $G \cup \partial G$ of the type described in Lemma 1.3. Clearly each function $h_i z \in \mathfrak{B}_\lambda$ on $G$ and on $R_i$ and the thansform $w_i(y)$ under $\tau_i \in \mathfrak{B}_\lambda$ on either $Q$ or $Q^+$; in the former case $w_i$ vanishes on and near $\partial Q$ and in the latter, $w_i$ vanishes near $\partial Q^+ \cap \cap \partial Q$. In the latter case, $w_i$ is equivalent to a function $w_{i0}$ which is $AC$ in $y^\alpha$ for almost all $y'_\alpha$, $\alpha = 1, \dots, \nu$ on any cell where $h \leq y^\nu \leq 1$ (since $w_i = 0$ near $y^\nu = 1$), where $h > 0$. But since $w_{i0,\nu} \in \mathfrak{L}_\lambda$, we see that $w_{i0}$ is $AC$ in $y^\nu$ for $0 \leq y^\nu \leq 1$ for almost all $y'_\nu$. If we extend $w_{i0}$ to the whole of $Q$ by setting

$$w_{i0}(y^\nu, y'_\nu) = + w_{i0}(-y^\nu, y'_\nu) \quad \text{for} \quad -1 \leq y^\nu \leq 0,$$

we see that $w_{i0} \in \mathfrak{B}'_\lambda$ on $Q$ and vanishes near $\partial Q$. Clearly we may approximate each $w_i$ or $w_{0i}$ on $Q$ strongly in $\mathfrak{B}_\lambda$ by functions $w_{ni}$ of class $C'$ on $\overline{Q}$ and vanishing near $\partial Q$. If we define $z_{ni}$ on $R_i$ as the transform of $w_{ni}$ under $\tau_i$ and then define $z_n = z_{n1} + \dots + z_{nN}$, we see that $z_n$ has the desired properties.

To prove (ii) we choose, in all cases, $w_{i0}$ equivalent to $w_i$ and $\mathfrak{B}'_\lambda$ on $Q$. Then, since $w_{i0}$ is $AC$ in $x^\nu$, we see that

$$\int_{-1}^{1} | w_{i0}(y_2^\nu, y'_2) - w_{i0}(y_1^\nu, y'_\nu) |^\lambda \, dy'_\nu$$

(1.7)

$$\leq (y_2^\nu - y_1^\nu)^{\lambda - 1} \int_{y_1^\nu}^{y_2^\nu} \int_{-1}^{1} | w_{i0,\nu}(y^\nu, y'_\nu) |^\lambda \, dy \leq \varepsilon \, (y_2^\nu - y_1^\nu),$$

$$0 < y_1^\nu < y_2^\nu; \ \lim_{\varrho \to 0^+} \varepsilon(\varrho) = 0.$$

Accordingly, we see that $w_{i0}(y^\nu, y'_\nu)$ converges strongly in $\mathcal{L}_\lambda$ in $y'_\nu$ to $w_{i0}(0, y'_\nu)$ as $y^\nu \to 0+$. If $z_n \to z$ in $\mathfrak{B}_\lambda$, $z_n$ of class $C'$, and we let $w_{ni}$ be the transform of $h_i z_n$ under $\tau_i$, then we see that (1.7) holds uniformly. Now let $\{p\}$ be any subsequence of $\{n\}$. There is a subsequence $\{q\}$ of $\{p\}$ such that (for each $i$) $w_{qi}(y^\nu, y'_\nu)$ converges strongly in $\mathcal{L}_\lambda$ with respect to $y'_\nu$ on $[-1, 1]$ for almost all $y^\nu$, $0 < y^\nu \le 1$. But, on account of the uniformity in (1.7), this convergence is uniform for *all* $y^\nu$, $0 \le y^\nu \le 1$. Hence the whole sequence $w_{i0}(0, y'_\nu)$ converges strongly to $w_{i0}(0, y'_\nu)$ in $\mathcal{L}_\lambda$.

(iii) is now evident. To prove (iiii), $\{\tilde{z}_n\}$ be of class $C'$ and converge strongly to $z$ in $\mathfrak{B}_\lambda$ on $G$. Then $\tilde{z}_n$ and each $h_i \tilde{z}_n$ converges strongly to 0 on $\partial G$. If we define the $w_{0i}$ as above, then $w_{0i}(0, y'_\nu) = 0$ for almost all $y'_\nu$ on $[+1, 1]$ if $R_i$ is a boundary neighborhood. If we extend such $w_{0i}$ to $Q$ by $w_{0i}(y^\nu, y'_\nu) = -w_{0i}(-y^\nu, y'_\nu)$, $y^\nu \le 0$ we see that $w_{0i}$ is of class $\mathfrak{B}_\lambda$ on $Q$ and that it and its $h$-average functions, for sufficiently small $h$ vanish near $\partial Q$ and along $y^\nu = 0$. By modifying the average function slightly for each $h$ in a sequence $\to 0$ we may construct sequences $w_{ni}$ tending strongly in $\mathfrak{B}_\lambda$ to $w_{0i} = 0$ such thot each $W_{ni} = 0$ near $y^\nu = 0$ as near $\partial Q$ for those $i$ for which $R_i$ is a boundary neighborhood. The desired $z_n$, each of class $C'$ and vanishing near $\partial G$ can be constructed as above.

THEOREM 1.15: *If $G$ is bounded and of class $C'$ and if $z_n \rightharpoonup z$ in $\mathfrak{B}_\lambda$ on $G$, then $z_n \to z$ in $\mathcal{L}_\lambda$ on $G$ and $\varphi_n \to \varphi$ in $\mathcal{L}_\lambda$ on $\partial G$. If $\| z_n \|$ is uniformly bounded in $\mathfrak{B}_\lambda$, and the set functions $\int_e |\nabla z_n| \, dx$ are uniformly absolutely continuous if $\lambda = 1$, there is a subsequence $\{z_p\}$ which converges weakly in $\mathfrak{B}_\lambda$ to some $z$ on $G$.*

*Proof:* Let $\{h_1, \ldots, h_N\}$, $w_{ni}$, $w_i$, and $w_{0i}$ have meanings as in the proof of Theorem 1.14 and let $w_{n0i}$ be of class $\mathfrak{B}'_\lambda$ on $Q$ (or $Q^+$) and be equivalent to $w_{ni}$ and extend each $w_{n0i}$ to $Q$ as before. Then (1.7) holds uniformly (in case $\lambda = 1$ this is true on account of the uniform absolute continuity in that case) and $w_{ni} \to w_i$ in $\mathcal{L}_\lambda$ on $Q$ for each $i$. The argument in the proof of (ii) in Theorem 1.14 can be repeated to obtain the desired results. The last statement follows easily.

In the next section, we shall have occasion to discuss vector functions of class $\mathfrak{B}_\lambda$.

DEFINITION: A vector function $z = (z^1, \ldots, z^N)$ is of class $\mathfrak{B}_\lambda$ if and only if each of its components is; in this case

$$\| z \|_{\beta_\lambda} = \left\{ \int_G \left[ \sum_{i=1}^{N} \left\{ (z^i)^2 + \sum_{\alpha=1}^{\nu} (z^i_\alpha)^2 \right\} \right]^{\lambda/2} dx \right\}^{1/\lambda}.$$

It is clear that all the theorems and lemmas of this section except Theorem 1.11 and lemma 1.2 generalize immediately to vector functions. Those two can be generalized with the help of the following well known lemma:

LEMMA 1.4: *Suppose $f_1, \dots, f_n$ are summable over the set $S$ with respect to the maesure $\mu$. Then $\sqrt{f_1^2 + \dots + f_n^2}$ is also and*

$$\left\{ \sum_{i=1}^{n} \left[ \int_S f_i(x)\, d\mu \right]^2 \right\}^{1/2} \leq \int_S \left[ \sum_{i=1}^{n} f_1^2(x) \right]^{1/2} dx . \tag{1.8}$$

*Proof:* For the left side of (1.8) equals

$$\max_{|a|=1} \sum_{i=1}^{n} \int_S a_i f_i(x)\, d\mu \leq \int_S \left[ \sum_{i=1}^{n} f_i^2(x) \right]^{1/2} d\mu ; \quad \left( |a|^2 = \sum_{i=1}^{n} a_1^2 \right).$$

In addition, we need the following special case of Rellich's theorem [53]:

THEOREM 1.16: *If the vector $z$ is of class $\mathcal{B}_\lambda$ on the hypercube $R$ of side $h$ and $z_R$ is its average over $R$, then*

$$\int_R |z(x) - z_R|^\lambda dx \leq C_2(\nu, \lambda) \cdot h^\lambda \cdot \int_R |\nabla z(x)|^\lambda dx$$

*where $C_2$ depends only on the arguments indicated.*

*Proof:* It is sufficient to prove this for vectors of class $C'$ where $R$: $|x^\alpha| \leq k = h/2$. Then we have

$$\int_R \int_R \left\{ \sum_{i=1}^{N} [z^i(\xi) - z^i(x)]^2 \right\}^{\lambda/2} dx\, d\xi$$

$$= \int_R \int_R \left\{ \sum_{i=1}^{N} \left[ \int_0^1 (\xi^\alpha - x^\alpha)\, z^i_{,\alpha}[x + t(\xi - x)]\, dt \right]^2 \right\}^{\lambda/2} dx\, d\xi$$

$$\leq \int_R \int_R \left[ \int_0^1 |\xi - x| \cdot |\nabla z[x + t(\xi - x)]|\, dt \right]^\lambda dx\, d\xi$$

$$\leq \int_{-k}^{k} \int_{-k}^{k} \int_0^1 |\xi - x|^\lambda \cdot |\nabla z[x + t(\xi - x)]|^\lambda dt\, dx\, d\xi .$$

**Setting $\eta^a = x^a + t(\xi^a - x^a) = (1-t)x^a + t\xi^a$, the last integral becomes**

$$\int_{-k}^{k}\left\{\int_{0}^{1} t^{-\nu-\lambda}\left[\int_{(1-t)x-tk}^{(1-t)x+tk} |\eta - z|^{\lambda}\cdot|\nabla z(\eta)|^{\lambda}\, d\eta\right] dt\right\} dx$$

$$= \int_{-k}^{k} |\nabla z(\eta)|^{\lambda}\left\{\int_{0}^{1} t^{-\nu-\lambda}\left[\int_{R(\eta,t)} |\eta - x|^{\lambda}\, dx\right] dt\right\} d\eta$$

**where $R(\eta, t)$ is the intersection of $R$ with the hypercube $|x^a - \eta^a/(1-t)| \leq tk$. On $R(\eta, t)$ we see that**

$$|\eta - x| \leq \nu^{1/2}\cdot th\,.$$

**The result follows since $m[R(h,t)] \leq h^{\nu}$ and is $\leq (2th)^{\nu}$ for $t \leq 1/2$.**

## CHAPTER II

### Lower-semicontinuity and existence theorems for a class of multiple integral problems.

In this chapter, we consider variational problems for integrals of the form (0,5) in which $f(x, z, p)$ is continuous in $(x, z, p)$ for all $(x, z, p)$ and is convex in $p$ for each $(x, z)$ (cf. [42], Chapter III).

DEFINITIONS: A set $S$ in a linear space is said to be *convex* if and only if the segment $P_1 P_2$ belongs to $S$ whenever the points $P_1$ and $P_2$ do. A function $\varphi(\xi)$ $(\xi = (\xi^1, \ldots, \xi^P))$ is said to be *convex* on the convex set $S$ in the $\xi$-space if and only if

$$\varphi[(1-\lambda)\xi_1 + \lambda\xi_2] \leq (1-\lambda)\varphi(\xi_1) + \lambda\varphi(\xi_2), \qquad 0 \leq \lambda \leq 1,$$

whenever $\xi_1$ and $\xi_2 \in S$.

The following theorems concerning convex functions are well known and are stated without proof:

LEMMA 2.1: *Suppose $\varphi(\xi)$ is convex on the open convex set $S$ with $|\varphi(\xi)| \leq M$ there. Then $\varphi$ satisfies*

$$|\varphi(\xi_2) - \varphi(\xi_1)| \leq 2M \cdot |\xi_2 - \xi_1| / \delta$$

*on any compact subset of $S$ at a distance $\geq \delta$ from $\partial S$.*

LEMMA 2.2: *Suppose $\varphi$ and each $\varphi_n$ are convex on the open convex set $S$ and suppose $\varphi_n(\xi) \to \varphi(\xi)$ for each $\xi$ on $S$. Then the convergence is uniform on any compact subset of $S$.*

LEMMA 2.3: *A necessary and sufficient condition that $\varphi$ be convex on the open convex set $S$ is that for each $\bar{\xi}$ in $S$ there exists a linear function $a_p \cdot \xi^p + b$ such that*

$$\varphi(\bar{\xi}) = a_p \bar{\xi}^p + b, \quad \varphi(\xi) \geq a_p \xi^p + b \qquad \text{for all} \qquad \xi \in S. \tag{2.1}$$

*If $\varphi$ is of class $C'$ on $S$, this condition is equivalent to*

$$E(\xi, \bar{\xi}) = \varphi(\xi) - \varphi(\bar{\xi}) - (\xi^a - \bar{\xi}^a)\varphi_{a'}(\bar{\xi}) \geq 0 ; \ \xi, \bar{\xi} \in S.$$

*If $\varphi$ is of class $C''$ on $S$, this condition is equivalent to*

$$\varphi_{,\alpha\beta}(\bar{\xi})\,\eta^{\alpha}\,\eta^{\beta} \geq 0$$

*for all $\bar{\xi}$ on $S$ and all $\eta$.*

DEFINITION: A linear function $a_p \xi^p + b$ which satisfies (2.1) for some $\bar{\xi}$ is said to *be supporting to $\varphi$* at $\bar{\xi}$.

LEMMA 2.4: *Suppose $\varphi$ is convex for all $\xi$ and satisfies*

$$\lim_{|\xi| + \infty} \varphi(\xi)/|\xi| = +\infty . \tag{2.2}$$

*Then $\varphi$ takes on its minimum. Also, if $a_1, ..., a_p$ are any numbers, there is a unique $b$ such that $a_p \xi^p + b$ is supporting to $\varphi$ for some $\bar{\xi}$. If $\psi$ is convex and satisfies (2.2), if $\psi(\xi) \geq \varphi(\xi)$ for each $\xi$, and if $a_p \xi^p + c$ is supporting to $\psi$, then $c \geq b$.*

LEMMA 2.5: *Suppose that $\varphi_n$ and $\varphi$ are everywhere convex and satisfy (2.2) and suppose that $\varphi_n(\xi) \to \varphi(\xi)$ for each $\xi$. Suppose $a_1, ..., a_p$ are any numbers and $b_n$ and $b$ are chosen so that $a_p \xi^p + b_n$ and $a_p \xi^p + b$ are supporting to $\varphi_n$ and $\varphi$, respectively. Then $b_n \to b$. Likewise, if $a_{np} \to a_p$ for each $p$ and $b_n$ and $b$ are chosen so that $a_{np} \xi^p + b_n$ and $a_p \xi^p + b$ are all supporting to $f$, then $b_n \to b$.*

In order to consider variational problems on arbitrary bounded domains, it is convenient to introduce the following type of weaker than weak convergence in $\mathfrak{B}_1$ on such a domain.

DEFINITION: We say that $z_n \dot{\rightharpoonup} z_0$ in $\mathfrak{B}_1$ on the bounded domain $G$ if and only if $z_n$ and $z_0$ all $\in \mathfrak{B}_1$ on $G$, $z_n \rightharpoonup z_0$ in $\mathfrak{B}_1$ on each cell interior to $G$ and each $z_{n,\alpha} \rightharpoonup z_{0,\alpha}$ in $\mathfrak{L}_1$ on the whole of $G$.

THEOREM 2.1: *If $G$ is bounded and of class $C'$ or if all the $z_n \in \mathfrak{B}_{10}$ on $G$ and if $z_n \dot{\rightharpoonup} z_0$ in $\mathfrak{B}_1$ on $G$, then $z_n \rightharpoonup z_0$ in $\mathfrak{B}_1$ on $G$.*

*Proof*: The second case can be reduced to the first by extending each $z_n$ to be zero outside $G$ and choosing a domain $\Gamma$ of class $C'$ such that $\Gamma \supset \bar{G}$. Thus we suppose $G$ of class $C'$. If we use the notation in the proof of Theorem 1.14, we see that (1.7) holde uniformly for the $w_{noi}$ so that an argument similar to those in the proofs of Theorems 1.14 and 1.15 and 1.13 shows that $w_{noi}$ converge strongly in $\mathfrak{L}_1$ on $Q$ or $Q^+$ to something for each $i$. Thus $z_n$ converges strongly in $\mathfrak{L}_1$ on $G$ to something which must be $z_0$.

REMARK: If $G$ is not of class $C'$ and the $z_n$ are not all in $\mathfrak{B}_{10}$ on $G$, then an example in [41] shows that $z_n \dot{\rightharpoonup} z_0$ in $\mathfrak{B}_1$ on $G$ without the $\mathfrak{B}_1$ norms of the $z_n$ being uniformly bounded. If for some $\lambda > 1$,

$$\int_G |\nabla z_n|^\lambda \, dz \equiv \int_G \left[\sum_{\alpha=1}^{\nu} z_{n,\alpha}^2\right]^{\lambda/2} dy \quad (G \text{ bounded})$$

are uniformly bounded, then a subsequence $\{p\}$ of $\{n\}$ exists such that the $z_{p,\alpha} \rightharpoonup$ something in $\mathfrak{L}_1$ on the whole of $G$.

THEOREM 2.2: *Suppose that $f(p)$ is defined of all $p = \{p_\alpha^i\}$ $(i = 1, \ldots, N$ $\alpha = 1, \ldots, \nu)$ and $f$ is convex. If $z_n \rightharpoonup z_0$ on $G$ and*

$$I(z_0, G) = \int_G f(\nabla z_0)\, dx, \quad I(z_n, G) = \int_G f(\nabla z_n)\, dx,$$

*then $I(z_0, G)$ and $I(z_n, G)$ are each finite or $+\infty$ and*

$$I(z_0, G) \leq \liminf_{n\to\infty} I(z_n, G).$$

*Proof*: Since $f$ is convex, there are constants $a_i^\alpha$ such that

$$f(p) \geq f(0) + a_i^\alpha p_\alpha^i \qquad \text{for all } p.$$

Hence

$$I(z, G) \geq f(0)\, m(G) + a_i^\alpha \int_G z_{,\alpha}^i (x)\, dx$$

with a similar inequality for $I(z_n)$. Thus the first statement follows.

If $D \subset G$, we see as above that

$$I(z_n, G) - I(z_n, D) = I(z_n, G - D) \geq f(0)\,[m(G - D)] + a_i^\alpha \int_{G-D} z_{n,\alpha}^i \, dx \geq$$
$$\geq \varepsilon\,[m(G - D)]; \quad \lim_{\varrho\to 0} \varepsilon(\varrho) = 0$$

by virtue of the uniform absolute continuity of the set functions $\int_e z_{n,\alpha}^i (x)\, dx$.

Clearly also $I(z, D) \to I(z, G)$ as $D$ runs through an expanding sequence of domains exhausting $G$. Thus it is sufficient to prove the lower semi-continuity for $G$ a hypercube of side $h$, say.

To do this, we define a sequence of summable functions $\varphi_q(x)$ as follows: For each $q$ divide $G$ into $2^{\nu q}$ hypercubes of side $h \cdot 2^{-q}$. On each of these hypercubes $R$, define

$$\varphi_q(x) = f(p_R) + a_i^\alpha(R, q)[z_{0,\alpha}^i(x) - p_{R\alpha}^i[, \quad x \text{ to interior to } R,$$

where $p_{R\alpha}^i$ is the average of $z_{,\alpha}^i$ over $R$ and the $a_i^\alpha(R, q)$ are chosen so that $f(p_R) + a_i^\alpha \cdot (p_\alpha^i - p_{R\alpha}^i)$ is supporting to $f$ at $p_R$. We define the $\varphi_{nq}$ similarly from $z_n$. Then it follows that

$$\varphi_q(x) \leq f[\nabla z_0(x)], \qquad \varphi_{nq}(x) \leq f[\nabla z_n(x)]$$

(almost everywhere). On the other hand, suppose all the generalized derivatives exist at some $x_0$ which is not on $\partial R$ for any hypercube $R$ as above for any $q$. Let $R$ denote the hypercube containing $x_0$. Then as $q \to \infty$ $p_{R\alpha}^i \to z_{,\alpha}^i(x_0)$ so that $\varphi_q(x_0) \to f[\nabla z(x_0)]$ since the $a_i^\alpha$ remain bounded (Lemma 2.5). Hence

$$I(z, G) = \lim_{q\to\infty} \int_G \varphi_q(x)\, dx. \tag{2.3}$$

Moreover, for each fixed $q$, $p_{nR} \to p_R$ from the weak convergence so

$$\int_G \varphi_q(x)\, dx = \sum_R f(p_R)\, m(R) = \lim_{n\to\infty} \sum_R f(p_{nR})\, m(R) =$$

$$= \lim_{n\to\infty} \int_G \varphi_{nq}(x)\ dx \leq \liminf_{n\to\infty} I(z_n, G).$$

The result follows from (2.3) and (2.4).

Lemma 2.6: *Suppose $f(x, z, p)$ is defined and satisfies a uniform Lipschitz condition with constant $K$ for all $(x, z, p)$, suppose $f(x, z, p)$ is convex in $p$ for each $(x, z)$ and suppose $f(x, z, p) \geq f_0(p)$ for all $(x, z, p)$, where $f_0(p)$ is convex. Then, if $z_n \rightharpoonup z_0$ in $\mathcal{B}_1$ on $G$,*

$$I(z_0, G) \leq \liminf_{n\to\infty} I(z_n, G).$$

*Proof:* As in the proof of Theorem 2.2, it is sufficient to prove this for a hypercube $D$ of side $d$ interior to $G$. Then $z_n \to z_0$ in $\mathcal{L}_\lambda$ on $D$. From the Lipschitz condition, $f(x, z, p) \leq f(0, 0, 0) + K \cdot |x| + K \cdot |z| + K \cdot |p|$ so that $I(z, D)$ and $I(z_n, D)$ are finite.

For each $q$, divide $D$ into $2^{\nu \cdot q}$ hypercubes $R$ of side $2^{-q} \cdot d$. Then, using Theorem 1.16, it follows that

$$\int_R |f[x, z(x), p(x)] - f[x_R, z_R, p(x)]| \, dx \leq K \int_R [|x - x_R| + |z(x) - z_R| \, dx \leq$$

$$\leq K \cdot 2^{-q} [2^{-1} \nu^{1/2} \cdot h^\nu + \int_R |\nabla z(x)| \, dx \qquad (h = 2^{-q})$$

$$\left| \int_D f[x, z(x), p(x)] \, dx - \sum_R \int_R f[x_R, z_R, p(x)] \, dx \right| \leq$$

$$\leq K \cdot 2^{-q} [2^{-1} \nu^{1/2} \cdot d^\nu + \int_D |\nabla z(x)| \, dx] \leq \varepsilon_q, \ \lim_{q \to \delta,} \varepsilon_q = 0$$

and a similar inequality holds for each $z_n$ with $\varepsilon_q$ independent of $n$ on account of the weak convergence. Also

$$\sum_R \int_R |f[x_R, z_{nR}, p_n(x)] - f[x_R, z_R, p_n(x)]| \, dx \leq K \int_D |z_n(x) - z(x)| \, dx.$$

The lemma follows easily from Theorem 2.2 and the inequalities above.

THEOREM 2.3: *Suppose $f(x, z, p)$ is defined and continuous for all $(x, z, p)$, is convex in $p$ for each $(x, z)$ and $f(x, z, p) \geq f_0(p)$ for all $(x, z, p)$ where $f_0(p)$ is convex and $f_0(p)/|p| \to +\infty$ as $p \to \infty$. Then $I(z, G)$ is lower semicontinuous with respect to the convergence $\dot{\to}$.*

*Proof:* In order to prove this, it is sufficient to show that $f(x, z, p)$ is the limit of a non-decreasing sequence $f_n(x, z, p)$ each of which has the properties required in Lemma 2.6. In order to do this, let $b(x, z; a)$ $(a = \{a_i^\alpha\})$ be chosen so that the function $\varphi(x, z; p; a) \equiv a_i^\alpha p^i + b(x, z; a)$ is the unique supporting plane (in $p$) to $f$ determined by $a$. By Lemmas 2.4 and 2.5 $b(x, z; a)$ is continuous in $(x, z; a)$ and $b(x, z; a) \geq b_0(a)$, the corresponding function for $f_0$. For each $a$, choose a non-decreasing sequence $b_n(x, z; a)$ of functions, each $\geq b_0(a) - 1$, each satisfyng a uniform Lipschitz condition for all $(x, z)$, which converges to $b(x, z; a)$. We then define $\varphi_n(x, z; p; a) = a_i^\alpha p_\alpha^i + b_n(x, z; a)$ and we see that $\varphi_n$ is a non decreasing sequence tending to $\varphi$ for each $a$, each $\varphi_n$ satisfying a uniform Lipschitz condition everywhere.

For each $n$, we define $f_n(x, z, p) = \max \varphi_n(x, z, p, a)$ for all a for which all the $a_i^\alpha$ are rational numbers having numerator and denominator both $\leq n$. Then it is clear that the $f_n$ are non-decreasing and each satisfies a uniform Lipschitz condition. Now, let $(x_0, z_0, p_0)$ and $\varepsilon > 0$ be given. Using Lemma 2.5 and the continuity of $b$, we see that there is a rational a such that $\varphi(z_0, p_0; \bar{a}) > f(x_0, z_0, p_0) - \varepsilon/2$. Clearly $\varphi_n(x_0, z_0, p_0; \bar{a}) > \varphi(x_0, z_0, p_0; \bar{a}) - \varepsilon/2$ for all sufficiently large $n$, so that $f_n(x_0, z_0, p_0) \to f(x_0, z_0, p_0)$.

We now turn to existence theorems on arbitrary domains. We begin with the following theorem (cf. [48] and [40], theorem 8.8 and [41]:

THEOREM 2.4: *Suppose $f_0(p)$ is convex in $p$ and $f_0(p)/|p| \to +\infty$ as $p \to \infty$. Then there is a function $\varphi(\varrho) \to 0$ as $\varrho \to 0$ which depends only on $f$ and $M$ such that if $I(z, G) \leq M$, then*

$$\int_e |\nabla z(x)| \, dx \leq \varphi[m(e)].$$

*Proof*: For each integer $r \geq 1$, let $E_r$ be the set of $x$ in $G$ where $r - 1 \leq |\nabla z(x)| < r$ and $\nabla z(x)$ exists and let

$$\mathcal{E}_r = \bigcup_{k=r+1}^{\infty} E_k \cup Z, \quad r = 0, 1, 2, \ldots$$

where $Z$ is the set of measure 0 where $\nabla z(x)$ does not exist. Clearly $\mathcal{E}_0 = G$ and if $r \geq 1$ and $x \in G - \mathcal{E}_r$, then $|\nabla z(x)| < r$. Let $\alpha_r$ be the inf. of $f_0(p)/|p|$ for $|p| \geq r - 1$. Then $\alpha_r \to +\infty$ as $r \to \infty$. Also

$$\sum_{k=r+1}^{\infty} \alpha_k \cdot (k-1) \cdot m(E_k) \leq \int_G f_0(\nabla z) \, dx \leq M$$

From this we see that

$$m(\mathcal{E}_r) \leq \frac{M}{r \cdot \alpha_{r+1}}, \quad \int_{\mathcal{E}_r} |\nabla z| \, dx \leq \frac{(r+1) M}{r \cdot \alpha_{r+1}}$$

and both $\to 0$ as $r \to \infty$. So, let e be any subset of $G$. Let $r$ be the smallest integer such that $M/r\,\alpha_{r+1} \leq m(e)$. Then

$$\int_e |\nabla z| \, dx \leq \int_{e - \mathcal{E}_r} |\nabla z| \, dx + \int_{e \cap \mathcal{E}_r} |\nabla z| \, dx$$

$$\leq \frac{M}{\alpha_{r+1}} + \frac{(r+1) M}{r \cdot \alpha_{r+1}} = \varphi[m(e)]$$

and $\varphi$ satisfies the conditions.

THEOREM 2.5: *Suppose $f(x, z, p)$ satisfies the hypotheses of Theorem 2.3 and $G$ is a bounded domain. Suppose that $\Gamma^*$ is a family of functions $z^*$ in $\mathfrak{B}_1$ which is compact with respect to the convergence $\dot{\rightarrow}$ in $\mathfrak{B}_1$ on $G$. Suppose $F$ is the family of all $z$ in $\mathfrak{B}_1$ which coincide on $\partial G$ in the $\mathfrak{B}_1$ sense with some $z^*$ in $\Gamma^*$ and suppose $F$ contains some $\tilde{z}_1$ for which $I(\tilde{z}_1, G) < +\infty$. Then $I(z, G)$ takes on its minimum in $F$.*

*Proof*: Let $\{z_n\}$ be a minimizing sequence (i. e. $I(z_n, G) \rightarrow$ greatest lower bound for $z$ in $F$); we may assume that $I(z_n, G) \leq M = I(\tilde{z}_1, G)$. Suppose $z_n = z_n^*$ on $G$ where $z_n^* \in \Gamma^*$. A subsequence $z_q^* \dot{\rightarrow} z_0^*$ in $\mathfrak{B}_1$ on $G$ and $z_0^* \in \Gamma^*$. By Theorem 2.4, the set functions $\int_e |\nabla z_q| \, dx$ are uniformly $AC$; the same is true of the set functions $\int_e |\nabla (z_q - z_q^*)| \, dx$. Since $G$ is bounded and each $z_q - z_q^* = 0$ on $\partial G$, we see with the aid of Theorem 1.13 that a subsequence $z_r - z_r^* \rightarrow$ some $w_0$ in $\mathfrak{B}_1$ ond $G$ and $w_0 = 0$ on $\partial G$. Accordingly $z_r \dot{\rightarrow} z_0 = z_0^* + w_0$ in $\mathfrak{B}_1$ on $G$ and $z_0 \in F$. the theorem follows from the lower-semicontinuity of $I(z, G)$.

Somewhat more meaningful boundary value problems can be studied if we require $G$ to be of class $C'$ at least. We need the following preliminary lemma:

LEMMA 2.7: *Suppose $G$ is bounded and of class $C'$ and $F$ is a family of functions of $\mathfrak{B}_\lambda$ on $G$ such that*

$$\int_G |\nabla z|^\lambda \, dx \leq M, \quad z \in F.$$

*Suppose that $F$ satisfies one of the following additional conditions:*

(i) *there is a number $P$ and an open subset $\tau$ of $G$ such that*

$$\int_\tau |z|^\lambda \, dx \leq P \text{ for all } z \in F; \text{ or}$$

(ii) *there is a number $P$ and an open set $\sigma$ of $\partial G$ such that*

$$\int_\sigma |z|^\lambda \, dS \leq P \text{ for all } z \in F.$$

*Then the $\mathfrak{B}_\lambda$ norms of the $z$ in $F$ are uniformly bounded.*

*Proof*: We may cover $G \cup \partial G$ with a finite number of hypercubes or boundary neighborhoods $R_1, \dots, R_Q$; let $\varphi_i$ map $Q$ or $Q^+$ onto $R_i$ as in the proof of Lemma 1.3. We may assume that one of the $R_i \subset \tau$ in case (i) or that $R_i \cap \partial G \subset \sigma$ in case (ii). In case (ii), we see using equation (1.7) with $y_1^\nu = 0$ that case (i) holds with $\tau = R_i$ and $P$ replaced by $P_i$; here we have assumed that $w_{i0}$ is equivalent to the transform under $\varphi_i$ of the restriction of $z$ to $R_i$.

Now, let $R_j \cap R_i$ be an open set $\tau_{ij}$. For a given $z$, let $w_{j0}$ be of class $\mathfrak{B}'_\lambda$ of $Q$ or $Q^+$ and be equivalent to the transform under $\varphi_j$ of the restriction of $z$ to $R_j$. Thus there is a cell $R_{j0} = [a, b]$ in $Q$ or $Q^+$ such that case (i) holds with $z$ replaced by $w_{j0}$ and $P$ by $P_{j0}$ (independently of $z$ in $F$). By using an equation like (1.7), we see in turn that case (i) holds with $R_{j0}$ replaced $R_{j1}, R_{j2}, \dots, R_{j\nu} = Q$ or $Q^+$ with $P$ replaced by $P_{j1}, \dots, P_{j\nu} = P'_j$ where $R_{j2}$ is the cell $-1 \leq x^1 \leq 1$, $-1 \leq x^2 \leq 1$, $a^\alpha \leq x^\alpha \leq b^\alpha$ for $\alpha = 3, \dots, \nu$, etc. Thus case (i) holds with $\tau$ replaced by $R_j$ and $P$ by $P_j$. Since any $R_k$ can be joined to the first $R_i$ by a sequence $R_n$, each two adjacent members of which have an open set in common, the lemma follows.

We can now prove our second principal existence theorem:

THEOREM 2.6: *Suppose the domain $G$ and the family $F$ satisfy the conditions of Lemma 2.7 for some $\lambda \geq 1$ and hence for $\lambda = 1$ and suppose $F$ contains some vector $\tilde{z}$ for which $I(\tilde{z}, G)$ is finite and suppose $F$ is closed with respect to weak convergence in $\mathfrak{B}_1$. Suppose that $f(x, z, p)$ satisfies the conditions of Theorem 2.3 Then $I(z, G)$ takes on its minimum in $F$.*

*Proof*: Let $\{z_n\}$ be a minimizing sequence for which $I(z_n, G) \leq I(\tilde{z}, G)$. Then the set functions $\int_e z_{n,\alpha}\, dx$ are uniformly absolutely continuous on account of Theorem 2.4. Combining this with Theorem 1.15; we see that a subsequence $\{z_p\}$ can be selected which converges weakly on $G$ in $\mathfrak{B}_1$ to some $z_0$ in $\mathfrak{B}_1$. Since $F$ is closed with respect to weak convergence in $\mathfrak{B}_1$, $z_0 \in F$. The result follows from the lower semicontinuity of $I(z, G)$.

THEOREM 2.7: *Suppose $G$ is of class $C'$, $f(x, z, p)$ satisfies the hypotheses of Theorem 2.3, and $\Gamma$ is a closed family of functions $\varphi$ in $\mathfrak{L}_1$ on $\partial G$ such that case* (ii) *of Lemma 2.7 holds. Suppose $F$ is the family of all functions $z$ in $\mathfrak{B}_1$ on $G$, each of which has boundary values in $\Gamma$ and suppose $F$ contains a function $\tilde{z}$ such that $I(\tilde{z}, G)$ is finite. Then $I(z, G)$ takes on its minimum in $F$.*

*Proof*: For the subfamily $\tilde{F}$ of $z$ in $F$ for which $I(z, G) \leq I(\tilde{z}, G)$ satisfies the conditions of Theorem 2.6, on account of Theorems 2.1, 2.3; and 1.15.

EXAMPLE: As an example of the use of Theorem 2.7, consider the problem of finding the surface $z = z(x)$ $(z = (z^1, z^2, z^3), x = (x^1, x^2))$ of least area of type of a disc bounded by a simple closed $C$ consisting of a fixed arc $C_1$ which has only its end points on a surface $S$ and a variable arc $C_2$ on $S$. Using theorems about conformal mapping this probleme can be reduced to that of minimizing the Dirichlet integral

$$I(z, G) = \int_G |\nabla z|^2 dx \left( \iint_G \sum_{i=1}^{3} (z^i_{x^1})^2 + (z^i_{x^2})^2 | dx^1 dx^2 \right)$$

among all vectors $z$ of class $\mathcal{B}_2''$ on $G$, where $G$ is the unit circular disc, such that the restrictions of $z$ to $\partial G$ carry the upper semicircle of $\partial G$ in a $1-1$ continuous way onto the fixed arc $C_1$ with $(0,1)$ corresponding to some fixed point on $C_1$ and carry the lower part of $\partial G$ in a $1-1$ continuous way onto the variable arc $C_2$. In order to apply Theorem 2.7, we let $\Gamma$ consist of all strong limits in $\mathcal{L}_2$ on $\partial G$ of the restrictions of such $z$ to $\partial G$. Any vector $\varphi$ in $\Gamma$ is equivalent along the upper part of $\partial G$ to a vector which carries that part of $\partial G$ in a « monotone » way onto $C_1$ in which arcs of $C_1$ may correspond to points on $\partial G$; for almost all $x$ on the lower part of $\partial G$, $\varphi(x) \in S$ at any rate. Since any minimizing vector $z_0$ certainly minimizes $I(z, G)$ among all $z$ in $\mathcal{B}_2$ which coincide with $z_0$ on $\partial G$ in the $\mathcal{B}_2$ sense, we see that $z_0$ is harmonic (see Professor Nirenberg's lectures). By arguments like those in [7] and [43], we conclude that $z_0$ is continuous on the upper half of $\partial G$ and yields a conformal map of $G$ onto the surface represented by $z_0$. However, an example of Courant [8] (p. 220, 221), shows that $z_0$ need not be continuous along the lower half of $\partial G$ and that the limiting « curve » $C_2$ need not be an arc even if the surface $S$ is regular and of class $C^\infty$; Lewy [33] has shown that if $S$ is analytic, the curve $C_2$ is analytic.

## Chapter III

## Quasi-convexity and lower-semicontinuity.

In the preceding chapter, we proved theorems concerning the lower-semicontinuity of multiple integrals $I(z, G)$ in cases where the integral function $f(x, z, p)$ is continuous and convex in $p$ for each $(x, z)$. This restriction on $f$ was a natural extension to the case of several unknown functions of the ordinary requirement when $N = 1$ that the variational problem be regular or at least that Hadamard's condition

$$(3.1) \qquad f_{p_\alpha p_\beta}(x, z\, p)\, \lambda_\alpha \lambda_\beta \geq 0 \qquad \text{for all } (x, z, p, \lambda)$$

be satisfied, $f$ being assumed of class $C''$. But (3.1) holds if and only if $f$ is convex in $(p_1, \ldots, p_\nu)$ for each $(x^1, \ldots, x^\nu, z)$.

The condition (3.1) is arrived at as follows: Suppose a function $z_0(x)$ of class $C'$ minimizes $I(z, G)$ among all functions of $z$ of class $C'$ which have the same boundary values and which are near $z_0$ in the sense that the maximum of $|z(x) - z_0(x)| + |\nabla z(x) - \nabla z_0(x)| \leq \delta$ for some $\delta > 0$. Then it can be shown that (3.1) holds for $x$ on $G$, $z = z_0(x)$, and $p = \nabla z_0(x)$. However, if this procedure is applied in the case where $N > 1$, we obtain only the condition

$$(3.2) \qquad f_{p_\alpha^j p_\beta^k}(x, z, p)\, \lambda_\alpha \lambda_\beta \xi^j \xi^k \geq 0$$

for all $(x, z, p)$ (along the solution $z = z_0(x)$, etc.) and all $(\lambda_1, \ldots, \lambda_\nu)$ and $(\xi^1, \ldots, \xi^N)$ (see Theorem 3.3 below). This does not imply that $f(x, z, p)$ is convex in $p$. Moreover, it is known that integrals $I(z, G)$ which arise in parametric problems are lower semi-continuous with respect to uniform convergence; for the case of the parametric problem for surfaces in 3-space $(\nu = 2, N = 3)$, these integrands have the form

$$f(x, z, p) = F(x, z, J_1, J_2, J_3)$$

where

$$J_1 = p_1^2 p_2^3 - p_1^3 p_2^2, \quad J_2 = p_1^3 p_2^1 - p_1^1 p_2^3, \quad J_3 = p_1^1 p_2^2 - p_1^2 p_2^1$$

and $F$ is convex in $(J_1, J_2, J_3)$, but not in the six $p_\alpha^i$.

It turns out to be rather easy to derive (see also [44]) a certain necessary and sufficient condition on $f$ as a function of $p$ for the lower semicontinuity of $I(z, G)$ with respect to a certain type of convergence. This question was considered for $\nu = N = 1$ by Tonelli ([72], [73], [74], [75]) and by Cesari and others for the parametric case. We begin by deriving this condition and then discuss the relation of that condition to the condition (3.2). In order not to get involved with the behavior of $f$ at infinity we shall use the following convergence which obviously implies weak convergence in each $\mathcal{B}_\lambda$ but does not necessarily imply strong convergence in any $\mathcal{B}_\lambda$:

DEFINITION: We say that $z_n \to z$ on $G \longleftrightarrow z_n(x)$ converges uniformly to $z(x)$ on $G$ and $z$ and $z_n$ each satisfy a uniform Lipschitz condition on $G$ which is independent of $n$.

THEOREM 3.1: *Suppose $I(z, G)$ is lower-semicontinuous with respect to this type of convergence at any $z$ on any $G$ and $f$ is continuous. Then.*

$$\int_0 f[x_0, z_0, p_0 + \nabla \zeta(x)]\, dx \geq f(x_0, z_0, p_0) \cdot m(G) \tag{3.3}$$

*for any constant $(x_0, z_0, p_0)$, any bounded domain $G$, and any Lipschitz vector $\zeta$ which vanishes on $\partial G$.*

*Proof:* Let $x_0$ be any point, $R$ be the cell $x_0^\alpha \leq x^\alpha \leq x_0^\alpha + h$, $z_0$ be any vector of class $C'$ on $R \cup \partial R$, $Q$ be the cell $0 \leq x^\alpha \leq 1$, and $\zeta$ be any vector which satisfies a uniform Lipschitz condition over the whole space and is periodic of period 1 in each $x^\alpha$.

For each $n$, define $\zeta_n(x)$ on $R$ by

$$\zeta_n^j(x) = n^{-1} h\, \zeta^j [n h^{-1} (x - x_0)].$$

Then the $\zeta_n^j$ tend to zero in our sense. Then, for each $n$, $I(z_0 + \zeta_n, R)$ can be written as a sum of integrals over the sub-hypercubes of $R$ of side $n^{-1} h$. If $r$ is one these the integral over it is

$$n^{-\nu} h^\nu \int_Q f[x_1 + n^{-1} h \xi, z_n(x_1 + n^{-1} h \xi), p_0(x_1 + n^{-1} h \xi) + \nabla \zeta(\xi)] d\xi,$$

where

$$r: x_1^\alpha \leq x^\alpha \leq x_1^\alpha + n^{-1} h, \; x_1^\alpha = x_0^\alpha + k^\alpha n^{-1} h, \; 0 \leq k^\alpha \leq n - 1$$

$$z_n(x) = z_0(x) + \zeta_n(x), \; x^\alpha = x_1^\alpha + n^{-1} h \xi^\alpha, \; 0 \leq \alpha \leq 1.$$

Thus we see that

$$\lim_{n\to\infty} I(z_0+\zeta_n, R) = \int_R \left\{ \int_Q f[x, z_0(x), p_0(x) + \nabla \zeta(\xi)]\, d\xi \right\} dx \geq I(z_0, R).$$

By letting $z_0$ and $p_0$ be arbitrary constant verctors, setting $z_0(x) = z_0 + p_{0\alpha} \cdot \cdot (x^\alpha - x_0^\alpha)$, dividing by $m(R) = h^\nu$ and letting $h \to 0$, we obtain (3·3) for $G = Q$ and $\zeta$ periodic of period 1 in each $x^\alpha$. But if $G$ is any bounded domain and $\zeta$ vanishes on $\partial G$, we may choose a hypercube $Q'$ containing $G$ and extend $\zeta(x)$ to be zero in $Q' - G$. Then a simple change of variable obtains the result in general.

DEFINITION: If $f$ is continuous in $(x, z, p)$ for all $(x, z, p)$ and satisfies (3.3) for all $(x_0, z_0, p_0)$, we say that $f$ is *quasi-convex in* $p$; if $f$ depends only on $p$ and satisfies (3.3), we say simply that $f$ is quasi-convex.

We now prove that the condition (3.3) is sufficient for lower-semicontinuity.

LEMMA 3.1: *Suppose* $R$ *is the hypercube* $|x^\alpha - x_0^\alpha| \leq h$, $f(p)$ *is quasi-convex, suppose* $p_0$ *is any constant tensor and suppose* $\zeta_n \to 0$ *in our sense or* $R$. *Then*

$$\liminf_{n\to\infty} \int_R f[p_0 + \nabla \zeta_n(x)]\, dx \geq f(p_0) \cdot m(R).$$

*Proof*: Suppose the $\zeta_n$ satisfy a uniform Lipschitz condition with constant $M$ on $R$. We may assume that $|\zeta_n(x)| \leq M k_n h$ where each $k_n < 1/2$ and $\lim k_n = 0$. For each $n$, we begin by defining $\eta_n(x) = \zeta_n(x)$ on $\partial R$ and $\eta_n(x) = 0$ for $|x^\alpha - x_0^\alpha| \leq (1 - k_n) h$; we then extend each $\eta_n$ to the whole of $R$ to satisfy a Lipschitz condition with constant $\leq M$. Then $\eta_n \to 0$, $\zeta_n - \eta_n \to 0$, $\zeta_n(x) - \eta_n(x) = 0$ on $\partial R$, and $\eta^j_{n,\alpha}(x) \to 0$ for each $x$ interior to $R$. Henee

$$\lim_{n\to\infty} \int_R |f[p_0 + \nabla \zeta_n] - f[p_0 + \nabla(\zeta_n - \eta_n)]|\, dx = 0.$$

The result follows easily from the quasi-convexity of $f$.

THEOREM 3.2: *Suppose* $f(x, z, p)$ *is quasi-convex in* $p$, $G$ *is a bounded domain, and* $z_n \to z_0$ on $G$. *Then*

$$I(z_0, G) \leq \liminf_{n\to\infty} I(z_n, G).$$

*Proof:* Since all the arguments $[x, z_n(x), \nabla z_n(x)]$ and $[x, z_0(x), \nabla z_0(x)]$ remain in a bounded part $\mathcal{G}$ of $(x, y, p)$-space and since $G$ is the union of $\mathcal{H}_0$ disjoint hypercubes, it is sufficient to prove this for the case of a hypercube $R$ of side $h$. Since $f$ is uniformly continuous on $\mathcal{G}$, there is a function $\varepsilon(\varrho)$ with $\lim_{\varrho\to 0} \varepsilon(\varrho) = 0$ such that

$$|f(x', z', p') - f(x'', z'', p'') \leq \varepsilon(\varrho) \quad \text{if} \quad |x' - x''|^2 + |z' - z''|^2 + |p' - p''|^2 \leq \varrho^2.$$

For each $k$, divide $R$ up into $2^{\nu k}$ hypercubes $R_{ki}$ of side $2^{-k} \cdot h$. Define the functions $x_k^*(x)$, $z_k^*(x)$, $p_k^*(x)$ on $R$ to be equal on each $R_{ki}$ to the averages over $R_{ki}$ of $x$, $z_0(x)$, and $p_0(x)$ respectively, and define

$$r_k(x) = \{ | x - x_k^*(x) |^2 + | z_0(x) - z_k^*(x) |^2 + | p_0(x) - p_k^*(x) |^2\}^{1/2}.$$

$$\zeta_n(x) = z_n(x) - z_0(x).$$

Then

$$f[x, z_n(x), \nabla z_n(x)] - f[x, z_0(x), \nabla z_0(x)] = A_n + B_{nk} - C_k + D_{nk} \tag{3.4}$$

where

$$\begin{aligned} A_n &= f[x, z_n(x), p_n(x)] - f[x, z_0(x), p_n(x)]; \quad (p_n(x) = \nabla z_n(x)) \\ B_{nk} &= f[x, z_0(x), p_0(x) + \pi_n(x)] - f[x_k^*(x), z_k^*(x), p_k^*(x) + \pi_n(x)] \\ C_k &= f[x, z_0(x), p_0(x)] - f[x_k^*(x), z_k^*(x), p_k^*(x)]; \quad (\pi_n(x) = \nabla \zeta_n(x)) \\ D_{nk} &= f[x_k^*(x), z_k^*(x), p_k^*(x) + \pi_n(x)] - f[x_k^*(x), z_k^*(x), p_k^*(x)]. \end{aligned} \tag{3.5}$$

We see that

$$\begin{aligned} | A_n | &\leq \varepsilon(| z_n(x) - z_0(x) |) \\ | B_{nk} |, \quad | C_k | &\leq \varepsilon[r_k(x)] \end{aligned} \tag{3.6}$$

and $I(z_n, R) - I(z_0, R) = J_n + K_{nk} + L_k + P_{nk}$, where these are the integrals of $A_n$, $B_{nk}$, $C_k$, and $D_{kn}$, respectively. Now, let $\varepsilon > 0$. We first choose a fixed $k$ such that $K_{nk}$ and $L_k$ are both $< \varepsilon/2$. From (3.4), (3.5), (3.6), and Lemma 3.1, we see that

$$\lim_{n\to\infty} J_n = 0, \quad \liminf_{n\to\infty} P_{nk} \geqq 0$$

since $x_k^*(x)$, $z_k^*(x)$, and $p_k^*(x)$ are each constant on each $R_{ki}$. Thus

$$\liminf_{n\to\infty}[I(z_n, R) - I(z_0, R)] \geq -\varepsilon .$$

Some of the theory of Chapter 2 can be carried over for the more general functions $f(x, z, p)$ which are quasi-convex in $p$ but more has to be assumed about how $f$ behaves as $p \to \infty$. These theorems are not of great interest and they can be found in [44].

We now investigate the concept of quasi-convexity in more detail.

LEMMA 3.2 [79], [45]: *Suppose $a_{jk}^{\alpha\beta}$ are constants and*

$$\int_G a_{jk}^{\alpha\beta} \zeta_{,\alpha}^j(x) \zeta_{,\beta}^k(x)\, dx \geq 0$$

*for all $\zeta$ in $\mathfrak{B}_{20}$ on domain $G$, then*

$$a_{jk}^{\alpha\beta} \lambda_\alpha \lambda_\beta \xi^j \xi^k \geq 0 \text{ for all } \lambda \text{ and } \xi. \tag{3.7}$$

*Proof:* Let $\lambda^1$ be a unit vector with $\lambda_\alpha^1 = \lambda_\alpha$ and choose $\lambda^2, \ldots, \lambda^\nu$ so $(\lambda^1, \ldots, \lambda^\nu)$ form a normal orthogonal set. Suppose $x_0 \in G$ and let $y^\gamma = \lambda_\alpha^\gamma \cdot (x^\alpha - x_0^\alpha)$. Choose $h_0$ and $R > 0$ so that the set of all $x$ for which $|y^1| \leq h_0$ and $|y_1'| \leq R$ is in $G$. Let $\xi$ be an arbitrary vector and define

$$\zeta_h^j(x) = \xi^j \varphi_h(y^1) \cdot \psi(|y_1'|),$$

where

$$\varphi_h(y^1) = h - |y^1| \text{ if } |y^1| \leq h, \ \psi(r) = R - r \text{ if } 0 \leq r \leq R$$

and $\varphi_h$ and $\psi(|y_1'|) = 0$ otherwise. Then it is easy to see that

$$\lim_{h\to 0}(2h)^{-1} \cdot I(\zeta_h, G) = \Gamma_{\nu-1} R^\nu a_{jk}^{\alpha\beta} \lambda_\alpha \lambda_\beta \xi^j \xi^k / \nu(\nu - 1) \geq 0$$

which proves the lemma.

We now prove the theorem mentioned in the introduction to this chapter.

THEOREM 3.3: *Suppose $f(x, z, p)$ is of class $C''$ for all $(x, z, p)$ near the locus $S$ of all points $[x, z, (x), \nabla z_0(x)]$ for $x$ in $G$ and suppose $z_0(x)$ is of class $C'$ on $G \cup \partial G$ and minimizes $I(z, G)$ among all Lipschitz $z$ which coincide with $z_0$ on $\partial G$ and are such that $|z(x) - z_0(x)| + |\nabla z(x) - \nabla z_0(x)| \leq \delta$ for some $\delta > 0$. Then (3.2) holds for all $(x, z, p)$ on $S$.*

*Proof:* For, let $\zeta$ be any Lipschitz function vanishing vanishing on and near $\partial G$. Then $z_0 + \lambda\zeta$ is sufficiently near $z_0$ for all sufficiently small

$\lambda$. So if $\varphi(\lambda) = I(z_0 + \lambda\zeta)$, we must have

$$\varphi''(0) = \int_G f_{p^j_\alpha p^k_\beta}[x, z_0(x), p_0(x)]\, \zeta^j_{,\alpha}\, \zeta^k_{,\beta}\, dx \geq 0$$

By selecting any point $x_0$ in $G$ and proceeding as in the proof of Lemma 3.2 and then dividing by $[\Gamma_\nu R^\nu/\nu(\nu-1)]$, but letting $R$ and $h$ both $\to 0$ so that $h : R \to 0$, we obtain (3.2) at $[x_0, z(x_0), p(x_0)]$.

Using the result of Lemma 3.2 and the method of proof of Theorem 3.3, we conclude that if $f(p)$ is quasi-convex and of class $C''$, then (3.2) holds with $x$ and $z$ omitted. This result and the analogy with convex functions suggest the following theorem whichwe now prove.

THEOREM 3.4: *If $f(p)$ is quasi convex, then $f(p^j_\alpha + \lambda_\alpha \xi^j)$ is convex in $\lambda$ for each $p$ and $\xi$ and convex in $\xi$ for each $p$ and $\lambda$.*

*Proof:* If $f$ is quasi-convex, it is easy to see that its twiceiterated $h$-average function $f_{hh}$ is also quasi-convex and is of class $C''$ as well. Then any linear function furnishes an absolute minimum to $I_{hh}(z, G)$ among all Lipschitz functions with the same boundary values. Accordingly, by Theorem 3.3 we see that $f_{hh}$ satisfies (3.2). But then $f_{hh}$ has the convexity properties stated in the theorem. Since $f_{hh}$ converges uniformly to $f$ on any bounded part of space, the theorem follows.

DEFINITION: A function $f(p)$ which satisfies the conditions in Theorem 3.4 is said to be weakly *quasi-convex.*

REMARK: The principal problem, so far unsolved, is whether or not every weakly quasi-convex function is quasi-convex.

THEOREM 3.5: *If $f(p)$ is weakly quasi-convex, it satisfies a uniform Lipschitz condition on a bounded part of space. If $p$ is given, there are constants $A^\alpha_j$ such that*

$$f(p^j_\alpha + \lambda_\alpha \xi^j) \geq f(p^j_\alpha) + A^\alpha_j \lambda_\alpha \xi^j \text{ for all } \lambda, \xi. \tag{3.8}$$

*If $f$ is also of class $C'$, then $A^\alpha_j = f_{p^j_\alpha}(p)$. If $f$ is also of class $C''$ then* (3.2) *holds. If $f$ is continuous and if, for each $p$, constants $A^\alpha_j$ exist such that* (3.8) *holds, then $f$ is weakly quasi-convex.*

*Proof:* If $f$ is weakly quasi-convex, it is convex in each $p^j_\alpha$ separately. Hence, if $|f(p)| \leq M$ on some hypercube, any difference quotient of the form:

$$|[f(p^j_{2\alpha}) - f(p^j_{1\alpha})]/(p^j_{2\alpha} - p^j_{1\alpha})| \leq 2M/d, \; p^j_{1\alpha} < p^{j\times}_{2\alpha}$$

where $d$ is the smaller of $b^j_\alpha - p^j_{2\alpha}$ and $p^j_\alpha - a^j_\alpha$.

Next, $f_{hh}$ is still weakly quasi-convex and of class $C''$ so that (3.2) holds. Then, from the convexity in $\xi$ for each $\lambda$, for instance, (3.8) holds with $A^{\alpha}_{hh} = f_{hhp^i_\alpha}(p)$. Since $f$ satisfies a uniform Lipschitz condition near $p$, we see that the $A^{\alpha}_{hhj}$ are uniformly bounded as $h \to 0$ so a sequence of $h \to 0$ can be chosen so that all the $A^{\alpha}_{hhj}$ tend to limits. Clearly (3.8) holds in the limit. Since the unit vector in the $p^j_\alpha$ direction is of form $\lambda_\alpha \xi^j$, we see that $A^{\alpha}_j = f_{p^j_\alpha}$ if $f$ is of class $C'$. The last statement follows from theorems on convex functions.

We now define a sufficient condition for $f$ to be (strongly) quasi-convex.

THEOREM 3.6: *A sufficient condition for $f$ to be quasi-convex is that for each $p$ there exist alternating forms*

$$A^{\alpha_1 \dots \alpha_\mu}_{j_1 \dots j_\mu} \pi^{j_1}_{\alpha_1} \dots \pi^{j_\mu}_{\alpha_\mu}, \qquad \mu = 1, \dots, \nu$$

(in which the coefficients are 0 unless all the $\alpha_1 \dots \alpha_\mu$ are distinct and all the $j_1 \dots j_\mu$ are distinct and an interchange of two $\alpha$'s or two $j$'s changes the sign) *such that for all $\pi$ we have*

$$f(p+\pi) \geq f(p) + \sum_{\mu=1}^{\nu} A^{\alpha_1 \dots \alpha_\mu}_{j_1 \dots j_\mu} \pi^{j_1}_{\alpha_1} \dots \pi^{j_\mu}_{\alpha_\mu}. \tag{3.9}$$

*Proof:* For suppose $p$ is any constant tensor, $G$ is any bounded domain, and $\zeta$ is any Lipschitz vector which vanishes on $\partial G$. By extending $\zeta = 0$ outside $G$ and approximating to it on a larger domain $D$ with smooth boundary with functions of class $C''$ which vanish on and near $\partial D$ and using Stokes' theorem we see that the integral of the *sum* on the right in (3.9) is zero. We now exhibit two interesting cases where the weak quasi-convexity of $f$ implies its quasi-convexity.

THEOREM 3.7: *If $f(p)$ is weakly quasi-convex and*

$$f(p) = a^{\alpha\beta}_{jk} p^j_\alpha p^k_\beta$$

*then $f$ is quasi-convex* ([79], [45]).

*Proof:* For, if $\zeta$ is Lipschitz and vanishes on $\partial G$ (which may as well be assumed smooth), then

$$\int_G f[p + \nabla \zeta(x)]\, dx = f(p) \cdot m(G) + \int_G a^{\alpha\beta}_{jk} \zeta^j_{,\alpha}(x) \zeta^k_{,\beta}(x)\, dx$$

If we introduce Fourier transforms (see [79])

$$Z^i(y) = (2\pi)^{-\nu/2} \int_G e^{iy^\alpha x^\alpha} \zeta^j(x)\, dx$$

we see that

$$\int_G a^{\alpha\beta}_{jk} \zeta^j_{,\alpha} \zeta^k_{,\beta} dx = \int_{-\infty}^{\infty} a^{\alpha\beta}_{jk} y^\alpha y^\beta Z^j(y) \overline{z}^k(y)\, dy \geq 0$$

since the integrand is $\geq 0$ for each $y$.

THEOREM 3.8: *If $N = \nu + 1$ and*

$$f(p) = F(X_1, \ldots, X_{\nu+1})$$

*where $F$ is continuous and*

$$X_j = \mp \det M_j \ (j = 1, \ldots, \nu), \quad X_{\nu+1} = \det M_{\nu+1}$$

$$M_{\nu+1} = \| p^1_\alpha, \ldots, p^\nu_\alpha \|, \quad M_j = \| p^1_\alpha, \ldots, p^{j-1}_\alpha p^{\nu+1}_\alpha, p^{j+1}_\alpha, \ldots, p^\nu_\alpha \| .$$

*Then $f$ is quasi-convex in $p$ if and only if $F$ is convex in $(X_1, \ldots, X_{\nu+1})$.*

We omit the proof which is found in [44]; $F$ is there required to be homogeneous of the first degree in $X$ but this is not necessary in the proof.

## Chapter IV

## The differentiability of the solutions of certain variational problems with $\nu = 2$.

In this chapter we discuss the differentiability of the solutions of certain problems whose existence was proved in § 2. To save time, we shall not discuss the continuity on the boundary but shall consider only the differentiability on the interior. This work was first presented in [42], chapters 4,6, and 7 and was the culmination of a series of papers on this subject by Lichtenstein [34], [35], Hopf [27], and the writer [39]. Some of these results have recently been generalized by De Giorgi [10] and Nash [49]. Sigalov [61] announced results similar to those presented here.

We begin with the following lemma which has a proper generalization for all values of $\nu$ (see [42] and [47]):

Lemma 4.1: *Suppose a vector $z(x) \in \mathfrak{B}_2$ on a domain $G$ and suppose that*

$$(4.1) \qquad \int_{B(x_0,r)} |\nabla z|^2 \, dx \leq L^2 (r/a)^{2\lambda} \qquad \text{for} \quad 0 \leq r \leq a,$$

*whenever $B(x_0, a) \subset G$. Then*

$$(4.2) \quad | z(x_2) - z(x_1) | \leq C_1(\lambda) \cdot L \cdot (|x_1 - x_2|/a)^\lambda \quad \text{for} \quad 0 \leq | x_1 - x_2 | \leq a,$$

*where*

$$C_1(\lambda) = 2^{1-\lambda} \pi^{-1/2} \lambda^{-1}$$

*for every pair of points $(x_1, x_2)$ in $G$ such that every point on the segment joining them is at a distance $\geq a$ from $\partial G$.*

*Proof:* We note first that if $\xi$ is on the segment and $s \leq a$,

$$\int_{B(\xi,s)} | \nabla z(y) | \, dy \leq \pi^{1/2} L a^{-\lambda} s^{1+\lambda},$$

using the Schwarz inequality. Next we write

$$| z(x_2) - z(x_1) | \leq | z(x) - z(x_1) | + | z(x) - z(z_2) |$$

$$| z(x) - z(x_k) | = | (x^\alpha - x_k^\alpha) \int_0^1 z_{,\alpha} [x_k + t(x - x_k)] dt |$$

$$\leq r \int_0^1 | \nabla z [x_k + t(x - x_k)] | dt, \; r = | x_2 - x_1 |, \; k = 1, 2,$$

and then average with respect to $x$ over $B(\bar{x}, r/2)$, $\bar{x} = (x_1 + x_2)/2$. If for a given $t$, $0 < t < 1$, we set $y = x_k + t(x - x_k)$, then $y$ ranges over $B[(1-t) x_k + t\bar{x}, rt/2]$. Then

$$\int_{B(x_0, r)} | z(x) - z(x_k) | dx \leq r \int_0^1 t^{-2} \left[ \int | \nabla z(y) | dy \right] dt$$

from which the result follows.

NOTATION: If $z \in \mathcal{B}_2$ on $G$, we define $D(z, G) = \int_G | \nabla z |^2 dx$; this is called the *Dirichlet integral.*

LEMMA 4.2: *Suppose* $z \in \mathcal{B}_2$ *on* $B(x_0, a)$ *and suppose*

$$D[z, B(x_0, r)] \leq K \cdot D[Z_r, B(x_0, r)] + \psi(r), \; 0 < r \leq a \tag{4.3}$$

where

$$\int_0^a s^{-1} \psi(s) \, ds$$

*converges, for every function* $Z_r = z$ *on* $\partial B(x_0, r)$. *Then*

$$D[z, B(x_0 r)] \leq D[z, B(x_0, a)] (r/a)^{1/K} + K^{-1} r^{1/K} \int_0^a \varrho^{-1 - \frac{1}{K}} \psi(\varrho) \, d\varrho \tag{4.4}$$

*and the right side tends to zero with r.*

*Proof*: Let $\varphi(r) = D[z, B(x_0, r)]$. Then $\varphi$ is absolutely continuous. For almost all $r$, $z(r, \theta)$ is $AC$ in $\theta$ with $|z_\theta(r, \theta)|$ in $\mathcal{L}_2$. For such $r$, define

$$Z_r(\varrho, \theta) = \bar{z}(r) + (\varrho/r)[z(r, \theta) - \bar{z}(r)], \bar{z}(r) = \frac{1}{2\pi}\int_0^{2\pi} z(r, \theta)\, d\theta .$$

Using Fourier series, one easily sees that

$$\int_0^{2\pi} |z(r, \theta) - \bar{z}(r)|^2 d\theta \leq \int_0^{2\pi} |z_\theta(r, \theta)|^2 d\theta \leq r\varphi'(r) \tag{4.5}$$

By computing $D_2[Z_r, B(x_0, r)]$ and using (4.5) we see that

$$\varphi(r) \leq Kr\varphi'(r) + \psi(r) \tag{4.6}$$

from which (4.4) follows easily. In order to see that the right side of (4.4) tends to zero with $r$, we note that

$$r^{1/K}\int_r^a \varrho^{-1-1/K}\psi(\varrho)\, d\varrho \leq \int_r^{r^{1/2}} \varrho^{-1}\varphi(\varrho)\, d\varrho + r^{1/2K}\int_{r^{1/2}}^a \varrho^{-1}\varphi(\varrho)\, d\varrho .$$

THEOREM 4.1: *Suppose $f(x, z, p)$ is continuous for all $(x, z, p)$ and is convex in $p$ for each $(x, z)$, and suppose there are constants $m$, $M$, and $k$ such that*

$$m|p|^2 - k \leq f(x, z, p) \leq M|p|^2 + k, M \geq m \geq 0, \tag{4.7}$$

*for all $p$. Suppose $I(z_0, G)$ is finite, $G$ is a bounded domain, and $z_0$ minimizes $I(z, G)$ among all $z$ in $\mathfrak{B}_2$ coinciding with $z_0$ on $\partial G$. Then $z_0$ satisfies* (4.1) *and* (4.2) *on $G$ with*

$$\lambda = m/2M \text{ and } L^2 = D[z_0 B(x_0, a)] + 2k\pi a^2/M . \tag{4.8}$$

*Thus $z_0$ satisfies a uniform Hölder condition on each compact subset of $G$.*

*Proof:* Suppose $\overline{B(x_0, r)} \subset G$ and let $Z_r$ be any function in $\mathfrak{B}_2$ on $B(x_0, r)$ and coinciding with $z_0$ on $\partial B(x_0, r)$. Then, from (4.7)

$$mD[z_0, B_r] - k\pi r^2 \leq I(z_0, B_r) \leq I(Z_r, B_r) \leq MD(Z_r, B_r) + k\pi r^2$$

$$D(z_0, B_r) \leq \frac{M}{m} D(Z_r, B_r) + \frac{2k\pi}{m} r^2 \qquad (B_r = B(x_0, r)) .$$

The result follows from Lemma 4.2.

For the remainder of this section, we shall assume that $f(x, z, p)$ satisfies the following condition in addition to (4.7):

GENERAL ASSUMPTIONS: *We assume that $G$ is a bounded domain, $f$ satisfies the conditions of Theorem 4.1, and*

(i) *$f$ is of class $C''$ for all $(x, z, p)$*

(ii) *there are functions $m_1(R)$, $M_1(R)$, and $M_2(R)$ with $0 < m_1(R) \leq M_1(R)$ for all $R \geq 0$ such that*

$$m_1(R) |\pi|^2 \leq f_{p^j_\alpha p^k_\beta} \pi^j_\alpha \pi^k_\beta \leq M_1(R) |\pi|^2 \tag{4.9}$$

$$\sum_{j=1}^{N} \left\{ \sum_{k=1}^{N} \left[ |f_{z^j z^k}| + \sum_{\alpha=1}^{2} f^2_{p^j_\alpha z^k} \right] + \sum_{\alpha=1}^{2} \left[ |f_{z^j x^\alpha}| + \sum_{\beta=1}^{2} f^2_{p^j_\alpha x^\beta} \right] \right\} \leq M_2(R) \cdot |p|^2 \tag{4.10}$$

*for all $(x, z, p)$ such that $|x|^2 + |z|^2 \leq R^2$.*

THEOREM 4.2: *Suppose $f$ and $G$ satisfy the general assumptions, $z_0$ satisfies the continuity conclusions of Theorem 4.1, and $\zeta$ is any Lipschitz function on $G$ which vanishes on and near $\partial G$, and $\varphi(\lambda) = I(z_0 + \lambda\zeta)$. Then $\varphi'(0)$ exists and*

$$\varphi'(0) = \int_G \{ f_{z^j}[x, z_0(x), p_0(x)] \zeta^j(x) + f_{p^j_\alpha}[x, z_0(x), p_\alpha(x)] \pi^j_\alpha \} \, dx \tag{4.11}$$

*Proof:* Let $F$ be the compact support of $\zeta$. Since $z_0$ is continuous on $F$, $|x|^2 + |z_0(x)|^2 \leq R^2$, for some $R$, for all $x$ on $F$. Then, for almost all $x$ on $F$,

$$f[x, z_0(x) + \lambda\zeta(x), p_0(x) + \lambda\pi(x)] = f[x, z_0(x), p_0(x)] + \lambda [f_{z^j} \zeta^j +$$

$$+ f_{p^j_\alpha} \pi^j_\alpha] + \lambda^2 \{ A^{\alpha\beta}_{jk}(x, \lambda) \pi^j_\alpha \pi^k_\beta + 2B^\alpha_{jk}(x, \lambda) \pi^j_\alpha \zeta^k + C_{jk}(x, \lambda) \zeta^j \zeta^k \}$$

where, for instance,

$$A^{\alpha\beta}_{jk}(x, \lambda) = \int_0^1 (1 - t) f_{p^j_\alpha p^k_\beta} [x, z_0(x) + t\lambda\zeta(x), p_0(x) + t\lambda\pi(x)] \, dt.$$

Clearly all the $A^{\alpha\beta}_{jk}$, $B^\alpha_{jk}$, and $C_{jk}$ are measurable and we conclude also from the general assumptions and the Lipschitz character of $\zeta$ that

$$\varphi(\lambda) - \varphi(0) - \lambda \int_G (f_{z^j} \zeta^j + f_{p^j_\alpha} \pi^j_\alpha) \, dx = \lambda^2 K(\lambda) \tag{4.12}$$

where $K(\lambda)$ is uniformly bounded for $|\lambda| \leq 1$. The result follows.

**Definition**: If $\varphi(0) = 0$ for every $\zeta$ as in Theorem 4.2, we say that $z_0$ furnishes a *stationary value* to the integral $I(z, G)$.

**Corollary**: *If $f$, $G$, and $z_0$ satisfy the conditions of Theorem 4.2 and if $z_0$ minimizes $I(z, G)$ among all sufficiently near $z$ ($\mathcal{B}_2$ sense) having the same boundary values, then $z_0$ furnishes a stationary value to $I(z, G)$.*

In order to obtain further differentiability properties of the solutions $z_0$, we must consider the solutions $u$ of equations

$$(4.13)\quad \int_G [v^j_{,\alpha}(a^{\alpha\beta}_{jk} u^k_{,\beta} + b^\alpha_{jk} u^k + e^\alpha_j) + v^j(b^\alpha_{kj} u^k_{,\alpha} + c_{jk} u^k + f_j)]\, dx = 0, \quad v \in \mathcal{B}_{20}$$

where all the coefficients are measurable and satisfy

$$(4.14)\quad m_1 |\pi|^2 \leq a^{\alpha\beta}_{jk}(x)\, \pi^j_\alpha \pi^k_\beta \leq M_1 |\pi|^2 \quad \text{for all } \pi,$$

$$a^{\beta\alpha}_{kj} = a^{\alpha\beta}_{jk}, \quad x \in G$$

$$(4.15)\quad \int_{B(x_0,s)\cap G} (|b|^2 + |c| + |f|)\, dx \leq M_2^2 r^{2\lambda}, \quad e \in L_2, \quad 0 < m_1 \leq M_1.$$

We begin by considering the case where $b^\alpha_{jk} = c_{jk} = 0$ and set

$$I_0(u, v; G) = \int_G v^j_\alpha a^{\alpha\beta}_{jk} u^k_{,\beta}\, dx.$$

From our general assumptions, we see that

$$(4.16)\quad m_1 \int_G |\nabla u|^2\, dx \leq I_0(u, u; G) \leq M_1 \int_G |\nabla u|^2\, dx.$$

From this result and the Poincare inequality (Theorem 1.11), we see that the space $\mathcal{B}_{20}$ is a Hilbert space if we take $I_0(u, v; G)$ as an inner product and that the resulting norm is topologically equivalent to the original $\mathcal{B}_2$ norm on $\mathcal{B}_{20}$.

**Lemma 4.3**: *If $S$ is any set of finite measure, then*

$$\int_S |x - x_0|^{h-2}\, dx \leq 2\pi \cdot h^{-1} s^h, \quad 0 < h < 2, \quad \pi s^2 = m(S).$$

*Proof*: Obviously $\int_S |x - x_0|^{h-2}\, dx \leq \int_{B(x_0,s)} |x - x_0|^{h-2}\, dx$,

LEMMA 4.4: *Suppose* $u \in \mathfrak{B}_{20}$ *on* $G$, $f \in \mathfrak{L}_1$ *on* $G$, *and*

$$\int_{B(x_0,r)\cap G} |f(x)|\,dx \le L r^{2\lambda}$$

*for every circle* $B(x_0, r)$. *Then* $u \cdot f \in \mathfrak{L}_1$ *on* $G$ *and satisfies*

$$\int_{B(x_0,r)\cap G} |f(x)\cdot u(x)|\,dx \le C_1(\lambda, \mu)\cdot L \cdot \|\nabla u\|_{L_2}\cdot g^{\mu}\, r^{2\lambda-\mu}, \quad 0 < \mu < \lambda,$$

$$C_1(\lambda, \mu) = 2^{-1}\,\pi^{-1/2}\,\lambda^{1/2}\,\mu^{-1/2}\,(\lambda-\mu)^{-1/2}, \quad \pi g^2 = m(G);$$

*u and f may be tensors.*

*Proof*: The proof for the general vector $u$ in $\mathfrak{B}_{20}$ will follow from the result for class $C'$ which vanishes near $\partial G$. Let $x_1 \in G$ and suppose $\overline{G} \subset B(x_1, R)$ and extend $u$ to be zero outside $G$. Then if we set $v^j(r, \theta) = u^j(x_1^1 + r\cos\theta,\, x_1^2 + r\sin\theta)$, we see that

$$\begin{aligned} u^j(x_1) = v^j(0, \theta) &= -\frac{1}{2\pi}\int_0^R\int_0^{2\pi} v^j_r(r, \theta)\,dx\,d\theta \\ &= -\frac{1}{2\pi}\int_G |\xi - x_1|^{-2}(\xi^\alpha - x_1^\alpha)\,u^j_{,\alpha}(\xi)\,d\xi. \end{aligned} \tag{4.17}$$

Hence

$$\begin{aligned} &\int_{B(x_0,r)\cap G} |f(x)\cdot u(x)|\,dx \le \\ &\le \frac{1}{2\pi}\int_{B(x_0,r)\cap G}\int_G |f(x)|\cdot|\xi - x|^{-1}\cdot|\nabla u(\xi)|\,d\xi\,dx. \end{aligned} \tag{4.18}$$

Applying the Schwarz inequality judiciously to (4.18), we obtain

$$\begin{aligned} \int_{B(x_0,r)\cap G} |f(x)\cdot u(x)|\,dx &\le \frac{1}{2\pi}\left[\int_{B(x_0,r)\cap G}\int_G |\xi - x|^{2\mu-2}\cdot|f(x)|\,dx\,d\xi\right]^{1/2} \\ &\left[\int_{B(x_0,r)\cap G}\int_G |f(x)|\cdot|\xi - x|^{-2\mu}\cdot|\nabla u(\xi)|^2\,dx\,d\xi\right]^{1/2}. \end{aligned} \tag{4.19}$$

Using Lemma 4.3 we see that

$$\int_{B(x_0,r)\cap G}\int_G |\xi - x|^{2\mu-2}\,|f(x)|\,dx\,d\xi \le \pi\mu^{-1}\,g^{2\mu}\cdot Lr^{2\lambda}. \tag{4.20}$$

Next, define,

$$\varphi_\xi(\varrho) = \int_{B(\xi,\varrho)\cap B(x_0,r)\cap G} |f(x)|\,dx.$$

From our assumption on $f$, we see that

$$\varphi_\xi(\varrho) \le L\varrho^{2\lambda} \qquad \text{and } Lr^{2\lambda}.$$

Accordingly

$$\int_{B(x_0,r)\cap G} |\xi - x|^{-2\mu}\,|f(x)|\,dx = \int_0^\infty \varrho^{-2\mu}\,\varphi'_\xi(\varrho)\,d\varrho =$$

$$= \int_0^\infty 2\mu\,\varrho^{-2\mu-1}\,\varphi_\xi(\varrho)\,d\varrho \le L\lambda\,(\lambda-\mu)^{-1}\,r^{2\lambda-2\mu}$$

(4.21)

$$\int_{B(x_0,r)\cap G}\int_G |f(x)|\cdot|\xi - x|^{-2\mu}\,|\nabla u(\xi)|^2\,dx\,d\xi \le$$

$$\le L\lambda\,(\lambda-\mu)^{-1}\,r^{2\lambda-2\mu}\cdot\|\nabla u\|^2_{L^2}.$$

The result follows from (4.20) and (4.21).

LEMMA 4.5: *Suppose $u$ and $f$ satisfy the hypotheses of Lemma 4.4. Then $fu^2 \in \mathcal{L}_1$ on $G$ and*

$$\int_{B(x_0,r)\cap G} |f(x)|\cdot|u(x)|^2\,dx \le C_2(\lambda,\mu)\cdot L\cdot\|\Delta u\|^2_{L^2}\cdot g^{\mu}\cdot r^{2\lambda-\mu};\quad 0 \le \mu \le \lambda.$$

*Proof:* This follows from two applications of Lemma 4.4.

THEOREM 4.3: *There is an $a_0 > 0$ and depending only on $m_1$, $M_1$, $M_2$, and $\lambda$ such that if $0 < a \le a_0$ and $B(x_0, a) \subset G$, then*

$$I[u, u; B(x_0, a)] \ge \frac{m_1}{2}\,D[u, B(x_0, a)] \quad \text{for all } u \in \mathcal{B}_{20} \text{ on } B(x_0, a).$$

*Proof:* For

$$I[u, u; B(x_0, a)] = I_0[u, u; B(x_0, a)] + \int_{B(x_0,a)} (2b_{jk}^{a} u_a^j u^k + c_{jk} u^j u^k)\, dx$$

$$\geq D_2[u, B(x_0; a)] \cdot [m_1 - 2C_2^{1/2} M_2^{1/2} g^{\mu/2} a^{\lambda-\mu/2} - C_2 M_2 g^{\mu} a^{2\lambda-\mu}], \quad 0 < \mu < \lambda,$$

using Lemma 4.5 and the Schwarz inequality.

THEOREM 4.4: *If $0 < a \leq a_0$, $B(x_0, a) \subset G$, $b_{jk}^{a}$, $c_{jk}$, and $f$ saiisfy (4.15) and $e \in \mathfrak{L}_2$ on $B(x_0, a)$, there exists a unique $u$ in $\mathfrak{B}_{20}$ on $B(x_0, a)$ such that (4.13) holds for all $v \in \mathfrak{B}_{20}$ on $B(x_0, a)$. Moreover*

$$(4.22) \qquad B[u, B(x_0, a)] \leq 2m_1^{-1}[\| e \|_L + C_1(\lambda, \mu) \cdot M_2 \cdot a^{2\lambda}]^2, \quad 0 < \mu < \lambda.$$

*Proof:* From Theorem 4.3 and the Poincare inequality (Theorem 1.1i), we see that the space $\mathfrak{B}_{20}$ is a Hilbert space if we introduce $I(u, v)$ as inner product. Since the equation (4.13) ($G = B(x_0, a)$) can be written

$$(4.23) \qquad I(u, v) = L(v), \; L(v) = \int_{B(x_0,a)} (e_j^a v_{,a}^j + f_j v^j)\, dx$$

and since $L(v)$ is a linear functional, we see from Hilbert space theory that there is a unique $u$ in $\mathfrak{B}_{20}$ which satisfies the equation. If, now, we revert to $\{D[u, B(x_0, a)]\}^{1/2}$ as norm, we see from (4.23) and Lemma 4.4 that the norm of $L(v)$ is given by the bracket on the right in (4.22). The inequality (4.22) follows by comparing the $I$ and $D$ norm.

We can now prove the interior boundedness theorem:

THEOREM 4.5: *Suppose $u \in \mathfrak{L}_2$ on $B(x_0, a) \subset G$ where $0 < a \leq a_0$, $u \in \mathfrak{B}_2$ on $B(x_0, r)$ and (4.13) holds for each $v \in \mathfrak{B}_{20}$ on $B(x_0, r)$ for each $r$ with $0 < r \leq a$. Then*

$$\{D[u, B(x_0, r)]\}^{1/2} \leq C_3(m_1, M_1)\{\| e \| + C_1 L a^{2\lambda} + (a - r)^{-1} \| u \|\}$$

$$(C_1 = \min_{0<\mu<\lambda} C_1(\lambda, \mu))$$

*the norm being the $\mathfrak{L}_2$ norms.*

*Proof:* Let $h$ be a fixed function of class $C^\infty$ with $h(s) = 1$ for $s \leq 0$ and $h(s) = 0$ for $s \geq 1$ and $0 \leq h(s) \leq 1$. Choose $R$ so $r < R < a$ and define

$$\zeta(x) = h[(|x - x_0| - r)/(R - r)], \; v^j = \zeta^2 u^j, \; U^j = \zeta u^j.$$

Then $v$ and $U \in \mathcal{B}_{20}$ on $B(x_0, R)$. Substituting in (4.13), we obtain

$$0 = I[U, U; B(x_0, R) + \int_{B(x_0,R)} (\zeta e_j U^j_{,\alpha} + \zeta f_j U^j + \zeta\zeta_{,\alpha} e^\alpha_j u^j - a^{\alpha\beta}_{jk} \zeta_{,\alpha} \zeta_{,\beta} u^j u^k)\, dx$$

$$\geq \frac{m_1}{2} |||U|||^2 - |||U|||[\|e\| + C_1(\lambda, \mu) M_2 \cdot R^{2\lambda}] - h_1 \cdot (R-r)^{-1} \|e\| \cdot \|u\| -$$

$$- h_1^2 M_1 \cdot (R-r)^{-2} \|u\|^2$$

where $|||U|||$ is the $\mathcal{B}_{20} - D$-norm and $\|u\|$ is the $\mathcal{L}_2$ norm. Since (4.24) holds for all $R < a$, the result follows.

LEMMA 4.6: *If $u \in \mathcal{B}_2$ on $B(x_0, R)$, there is a $u_1 \in \mathcal{B}_{20}$ on $B(x_0, 2R)$ such that $u_1(x) = u(x)$ on $B(x_0, R)$ and*

$$D[u_1, B(x_0, 2R)] \leq C_4 \int_{B(x_0,R)} (|\nabla u|^2 + R^{-2}|u|^2)\, dx$$

*where $C_4$ is an absolute constant.*

*Proof:* Define $u_2(x) = u(x)$ on $B(x_0, R)$ and extend it by reflection in the circle $B(x_0, R)$. Then $u \in \mathcal{B}_2$ on $B(x_0, 2R)$ and

$$\int_{B(x_0,2R)-B(x_0,R)} |\nabla u_2|^2\, dx \leq \int_{B(x_0,R)} |\nabla u|^2\, dx$$

$$\int_{B(x_0 2R)-B(x_0,R)} |u_2|^2\, dx \leq 16 \int_{B(x_0,R)} |u|^2\, dx$$

Then, define

$$u_1(x) = h[(|x - x_0| - R)/R] \cdot u_2(x),$$

where $h$ is function introduced in the proof of Theorem 4.5. Then $u_1$ is easily seen to have the desired properties.

THEOREM 4.6 (Dirichlet growth theorem): *Suppose $0 < a \leq a_0$, $B(x_0, a) \subset G$, $u \in \mathcal{B}_2$ on $B(x_0, a)$, (4.13) holds for all $v \in \mathcal{B}_{20}$ on $B(x_0, a)$, and $e$ satisfies the condition*

$$\int_{B(x_1,r)} |e|^2\, dx \leq L^2 (r/\delta)^{2\mu}, \quad 0 \leq r \leq \delta = a - |x_1 - x_0|,$$

*for some $\mu$ with $0 < \mu < \lambda/2$ and $m_1/2M$, and every circle $B(x_1, r) \subset B(x_0, a)$. Then $u$ satisfies the condition (4.1) and (4.2) with $G$ replaced by $B(x_0, a)$, $x_0$*

*replaced by* $x_1$, $a$ *by* $\delta = a - |x_1 - x_0|$, $\lambda$ *replaced by* $\mu$, *and* $L$ *replaced by* $C_5$, *where* $C_5$ *depends only on* $m_1$, $M_1$, $M_2$, $L$, $\lambda$, $\mu$, $a$, *and* $|||u|||$ *where*

$$\int_{B(x_0,a)} (|\nabla u|^2 + a^{-2}|u|^2)\,dx \equiv |||u|||^2 .$$

*Thus* $u$ *satisfies a uniform Hölder condition on any* $B(x_0, R)$ *with* $R < a$ *which depends only on the quantities above and* $a - R$.

*Proof:* Let

$$E_j^\alpha = b_{jk}^\alpha u^k + e_j^\alpha, \quad F_j = b_{kj}^\alpha u_{,\alpha}^k + c_{jk} u^k + f_j .$$

From our hypotheses on the $b$'s, $c$'s, $e$'s, and $f$'s and from Lemmas 4.4, 4.5, and 4.6, we see that

$$\left\{ \int_{B(x_1,r)} |E|^2\,dx \right\}^{1/2} \leq C^{1/2}(\lambda, \mu') \cdot M_2 \cdot |||u||| \cdot a^{\mu'/2} r^{\lambda-\mu'/2} + L(r/\delta)^\mu \quad 0 < \mu' < \lambda ,$$

$$\int_{B(x_1,r)} |F|\,dx \leq [M_2 r^\lambda + C_1(\lambda, \mu') \cdot M_2 a^{\mu'} r^{2\lambda-\mu'}] \cdot |||u||| + M_2^2 r^{2\lambda} .$$

Moreover $u$ satisfies the equation

$$(4.25) \quad I_0[u, v; B(x_1, r)] = -\int_{B(x_1,r)} (E_j^\alpha v_{,\alpha}^j + F_j v^j)\,dx ; \quad v \in \mathfrak{B}_{20} \text{ on } B(x_1, r)$$

on any $B(x_1, r) \subset B(x_0, a)$. As in the proof of Theorem 4.4, there is a unique solution $U_r$ of (4.25) which is in $\mathfrak{B}_{20}$ on $B(x_1, r)$ and

$$D[U_r, B(x_1, r)] \leq m_1^{-1} [Z_1 a^\lambda |||u||| + L]^2 (r/\delta)^{2\mu}$$

where $Z_1$ depends only on the quantities mentioned.

Now $V_r = u - U_r$ satisfies the homogeneous equation (4.25) and so clearly minimizes $I_0[V, V; B(x_1 r)]$ among all $V = V_r (= u)$ on $\partial B(x_1, r)$. Since $U_r \in \mathfrak{B}_{20}$ on $B(x_1, r)$, we see that

$$I_0(V_r, U_r; B_r) = 0 \text{ so } I_0(u, u; B_r) = I_0(V_r, V_r; B_r) + I_0(U_r, U_r; B_r) ,$$

where $B_r = B(x_1, r)$. Using the fact that $I_0(V_r, V_r; B_r) \leq I_0(u_r, u_r; B_r)$ for any $u_r = u$ on $\partial B(x_1, r)$ and using (4.16), we see that

$$D(u, B_r) \leq \frac{M_1}{m_1} D(u_r, B_r) + Z_2 (r/\delta)^{2\mu}$$

where $Z_2$ depends only on the quantities indicated. The results follow from Lemmas 4.2 and 4.1.

We can now resume our discussion of a solution $z_0$ of a variational problem of the type being discussed here.

THEOREM 4.7: *Suppose $z_0$ gives a stationary value to $I(z, G)$ and satisfies the continuity conclusions of Theorem 4.1. Then $z_0 \in C^{1+\mu}$ on each domain $\Gamma$ with $\overline{\Gamma} \subset G$, where $0 < \mu < 1$, and the derivatives $\in \mathfrak{B}_2''$ on domains interior to $G$.*

*Proof*: Since $\varphi(0) = 0$, we see that the right side of (4.11) holds for each Lipschitz $\zeta$ with compact support in $G$. So, suppose $\overline{B(x_0, a)} \subset G$. Choose $A > a$ so that $\overline{B(x_0, A)} \subset G$. Then, from Theorem 4.1, we have $|x|^2 + |z_0(x)|^2 < R^2$, for some $R$, on $\overline{B(x_0, A)}$. Let $b = (2a + A)/3$, $c = (a + 2A/3)$, $h_0 = (A - a)/3$, let $e_\gamma$ be the unit vector in the $x^\gamma$ direction for $\gamma = 1,2$, let $v$ be an arbitrary Lipschitz function having support in $B(x_0 c)$ and define

$$\zeta_h^j(x) = h^{-1}[v^j(x - he_\gamma) - v^j(x)], \; u_h^j(x) = h^{-1}[z_0^j(x + he_\gamma) - z_0^j(x)]$$

for $0 < |h| < h_0$. Then $\zeta_h$ has support in $B(x_0, A)$. Substituting $\zeta_h$ into the equation $\varphi'(0) = 0$ and using (4.11), we see that $u_h$ satisfies equation (4.13) on $B(x_0, c)$ with coefficients $a_{hjk}^{\alpha\beta}$, etc., where

$$(4.26)\quad a_{hjk}^{\alpha\beta}(x) = \int_0^1 f_{p_\alpha^j p_\beta^k}[x + the_\gamma, (1-t)z_0(x) + tz_0(x + he_\gamma), (1-t)p_0(x) + tp_0(x + he_\gamma)]\,dt$$

for almost all $x$. From the general assumptions on $f$ and from the formulas (4.26) for the coefficients, we see that the bounds (4.14) and (4.15) hold uniformly for $0 < |h| < h_0$ with

$$m_1 = m_1(R), M_1 = M_1(R), M_2 = KM_2(R), 2\lambda = m/M, G = B(x_0, c),$$

where $K$ is a constant depending on $\lambda$ and the distance of $B(x_0, A)$ from $\partial G$. Clearly each $u_h \in \mathfrak{B}_2''$ on $B(x_0, c)$ and its $L_2$ norm is uniformly bounded there, and we also have

$$\int_{B(x_1, r)} |e_h|^2\,dx \leq M_2^2 r^{2\lambda}, \quad 0 < |h| < h_0.$$

Accordingly, we see first from Theorem 4.5 that the $\mathfrak{B}_2$ norms of the $u_h$ are uniformly bounded on $B(x_0, b)$ and then from Theorem 4.6 that the $u_h$ satisfy a uniform Hölder condition on $\overline{B(x_0, a)}$ independently of $h$. Thus we may let $h \to 0$ and we see that the derivatives $z_{,\gamma}^j \in \mathfrak{B}_2''$ and satisfy this Hölder condition on $\overline{B(x_0, a)}$.

## CHAPTER V

## A variational method in the theory of harmonic integrals.

In this section, we apply our variational method to the study of armonic integrals and, more generally, use it to obtain the Kodaira decomposition theorem [29] (see Theorem 5.10 below). This approach was originally suggested by Hodge in his first paper on the subject [25]. The generality of the manifolds allowed and the methods and results obtained are closely related to those obtained by Friedrichs [20] working independently. Of course corresponding results have been obtained on smoother manifolds by a number of other authors using other methods ([12], [23], [26], [29], [38]). In this section, we shall confine ourselves to compact manifolds without boundary. The variational methods are applied to compact manifolds with boundary in [20] and [46]; boundary value problems for forms have been considered by other writers using other methods in [13], [66].

We adopt the usual definition of a compact Riemannian manifold of dimension $n$ (instead of $\nu$) and of class $C^k$ or $C^k_\mu$ $(0 < \mu \leq 1)$ any two admissible coordinate systems are related by a transformation of class $C^k$ or $C^k_\mu$, respectively. If $0 < \mu < 1$, the class $C^k_\mu$ is the same as what we have called $C^{k+\mu}$; If $\mu = 1$, a function is of class $C^k_1$ if and only if its derivatives of order $\leq k$ satisfy Lipschitz conditions; transformations of class $C^k_1$ are defined similarly. If a coordinate system is of class $C^k_\mu$, the induced $g_{ij}$ are of class $C^{k-1}_\mu$. *We shall assume that our manifold is of class at least* $C^1_1$.

We shall be concerned with exterior differential forms of degree $r$ on a manifold $M$; we call these simply *r-forms*. In the domain of a given coordinate system such a form $\omega$ may be represented by

$$\omega = \sum_{i_1 < \dots < i_r} \omega_{i_1 \dots i_r} dx^{i_1} \wedge \dots \wedge dx^{i_r} \tag{5.1}$$

where $\omega_{i_1 \dots i_r}$ are the *components* of $\omega$ in that coordinate system and $\wedge$ denotes the exterior product. In order to take care of the case of non-orientable manifolds, we allow both *even* and *odd* forms. If two coordinate systems $(x)$ and $(x')$ overlap, the components transform according to the law

$$\begin{aligned} {}'\omega_{i_1 \dots i_r}(r') &= \varepsilon \sum_{j_1 < \dots < j_r} \omega_{j_1 \dots j_r}[x(x(x'))] \frac{\partial (x^{j_1} \dots x^{j_r})}{\partial ('x^{i_1} \dots 'x^{i_r})}, \\ \varepsilon &= \begin{cases} +1 \text{ for even forms,} \\ J/|J| \text{ form odd forms,} \end{cases} \quad J = \frac{\partial (x^i \dots x^n)}{\partial ('x^i, \dots, 'x^n)}. \end{aligned} \tag{5.2}$$

Since the Jacobians involved in (5.2) are at least of class $C_1^0$ (Lipschitz), we may say that a form $\omega$ *is of class* $\mathfrak{L}_2$ or $\mathfrak{B}_2 \longleftrightarrow$ its components in each coordinate system are.

Given an $r$-form $\omega$, we define its dual $*\omega$ by

$$*\omega = \sum_{j_1<\dots<j_{n-r}} (*\omega)_{j_1\dots j_{n-r}}\, dx^{j_1}\wedge\dots\wedge dx^{j_{n-r}}$$

$$(5.3)\qquad (*\omega)_{j_1\dots j_{n-r}} = \Gamma\cdot e_{k_1\dots k_r j_1\dots j_{n-r}} \sum_{l_1\dots l_r} g^{k_1 l_1}\dots g^{k_r l_r}\,\omega_{l_1 l_r}$$

$$= \Gamma\cdot e_{k_1\dots k_r j_1\dots j_{n-r}} \sum_{l_1<\dots<l_r} \Gamma^{(k)(l)}\,\omega_{l_1\dots l_r}$$

where $e_{p_1\dots p_n}$ is 0 if two indices $p_i$ are the same or otherwise is $\pm 1$ according as $p_1\dots p_n$ is an even or odd permutation, $k_1<\dots,<k_r$ are chosen so that $k_1\dots k_r j_1\dots j_{n-r}$ is a permutation, $\Gamma^{(k)(l)}$ is the determinant of the $g^{k_i l_j}$, and $\Gamma = \pm\sqrt{g}$ chosen so that $\Gamma\, dx^1\wedge\dots\wedge dx^n = dS$, the positive volume element. *If two forms* $\omega$ *and* $\eta$ *of the same kind* (both even or both odd) *of the same degree are in* $\mathfrak{L}_2$ *on* $M$, we define their *inner product*

$$(5.4)\qquad (\omega,\eta) = \int_M \omega\wedge *\eta\,;$$

we form inner products only under these conditions. If $\omega$ is an $r$-form given in the $x$-system by (5.1) and if $\eta$ is an $s$-form of the same kind with a corresponding representation, we define

$$(5.5)\qquad \omega\wedge\eta = \sum_{(i)}\sum_{(j)} \omega_{i_1\dots i_r}\,\eta_{j_1\dots j_s}\, dx^{i_1}\wedge\dots\wedge dx^{i_r}\wedge dx^{j_1}\wedge\dots\wedge dx^{j_s}.$$

Accordingly the inner product $(\omega,\eta)$ is also given by

$$(\omega,\eta) = \int_M (F(P;\omega,\eta)\, dS_P$$

$$(5.6)\qquad F = \sum_{(i)(j)} \Gamma^{(i)(j)}\,\omega_{(i)}\,\eta_{(j)}$$

where $(i) = i_1\dots i_r$, where $i_1<\dots<i_r$, etc. In case $P$ corresponds to $x_0$ in the $x$ system and $g_{ij}(x_0) = \delta_{ij}$, we see that

$$(5.7)\qquad F(P;\omega,\eta) = \sum_{(i)} \omega_{(i)}(x_0)\,\eta_{(i)}(x_0),\ dS_P\,|\,dx\,|\,.$$

The following theorem is well hoown and is evident.

THEOREM 5.1. *For each $r = 0, 1, \ldots, n$ the totality of r-forms of a fixed kind $\mathcal{L}_2$ on M (with equivalent forms identified) forms a real Hilbert space $\mathcal{L}_2^r$ with inner product given by* (5.4)

In order to introduce an inner product in $\mathcal{B}_2$ on $M$, we proceed as follows:

DEFINITION: Let $\mathcal{U} = (U_1, \ldots, U_Q)$ be a finite open covering of $M$ by coordinate patches $U_q = Q_q(G_q)$, where each $G_q$ is a Lipschitz domain in $\mathcal{E}^n$. If $\omega$ and $\eta$ are in $\mathcal{B}_2$ on $M$ we define

$$((\omega, \eta))_{\mathcal{U}} = (\omega, \eta) + \sum_{q=1}^{Q} \int_{G_q} \sum_{(i)} \sum_{\alpha=1}^{n} \omega^{(q)}_{(i)x^\alpha} \eta^{(q)}_{(i)x^\alpha} \, dx, \tag{5.8}$$

where $\omega^{(q)}_{(i)}$ and $\eta^{(q)}_{(i)}$ are the components of $\omega$ and $\eta$ in $Q_q$. Then

$$\| \omega \|_{\mathcal{U}} = ((\omega, \omega))^{1/2}_{u} \tag{5.9}$$

is the expression for the norm in $\mathcal{B}_2$ on $M$ corresponding to the inner product (5.8). It is char that convergence of $\omega_k$ to $\omega$ according to one of the norms (5.9) is equivalent to the atrong convergence in $\mathcal{B}_2$ of the compenents $\omega_k$ in *any* coordinate system to those of $\omega$. Thus we obtain the theorem:

THEOREM 5.2: *For each coordinate cover $\mathcal{U}$ and each $r = 0, \ldots, n$ the space of r-forms in $\mathcal{B}_2$ of a given kind on M forms a real Hilbert space $\mathcal{B}_2^r$ with inner product given by* (5.8). *Any two such inner product sare topologically equivalent.*

Now, if $\omega$ is an $r$ form $\in \mathcal{B}_2$, we define $d\omega$ and $\delta\omega$ by

$$\delta\omega = (-1)^{1+n(r-1)} * d * \omega, \text{ and}$$

$$d\omega = \sum_{(i)} \sum_{q=1} \omega_{i_1 \ldots i_r, q} \, dx^q \vee dx^{i_1} \vee \ldots \vee dx^{i_r}. \tag{5.10}$$

We note that $d\omega$ is an $(r+1)$-form (if $r \leq n-1$) and $\delta\omega$ is an $(r-1)$-form (if $r \geq 1$). Finally, we define the Dirichlet integral by

$$D(\omega) = (d\omega, d\omega) + (\delta\omega, \delta\omega). \tag{5.11}$$

THEOREM 5.3: *d is a bounded operator from the whole of $\mathcal{B}_2^r$ into $\mathcal{L}_2^{r+1}$, and $\delta$ is a bounded operalor from the while of $\mathcal{B}_2^r$ into $\mathcal{L}_2^{r-1}$; each of these operators preserves evenness or oddness. $D(\omega)$ is a lower semi-continuous function with respect to weak convergence in $\mathcal{B}_2^r$. If $\omega_k$ tends weakly to $\omega_0$ in $\mathcal{B}_2^r$ on M, then $\omega_k$ tends strongly to $\omega_0$ in $\mathcal{L}_2^r$ on M.*

*Proof.* The first statement in clear form (5.8) since the $g_{ij}$ are at least Lipschitz and have bounded first derivatives. New if $\omega_k$ tends weakly in $\mathcal{B}_2$ to $\omega$, $d\omega_k$ and $\delta\omega_k$ tend weakly in $\mathcal{L}_2$ to $d\omega$ and $\delta\omega$, whence the last statement about $D(\omega)$ follws from the lower-semicontinuity of the norm in $\mathcal{L}_2$ with respect to weak convergence. The last statement is an application of Theorem 1.13.

From (5.6) and (5.7), we see that

$$(\omega, \eta) = (\eta, \omega). \tag{5.12}$$

In the coordinate system of (5.7), we see that

$$(*\omega)_{j_1 \dots j_{n-r}} = e_{i_1 \dots i_r j_1 \dots j_{n-r}} \, \omega_{i_1 \dots i_r} \qquad ((i) \text{ not summed}) \tag{5.13}$$

where $i_1 < \dots < i_r$ and $i_1 \dots i_r j_1 \dots j_{n-r}$ is a permutation. From the form (5.12), we see that

$$** \omega = (-1)^{r(n-r)} \omega. \tag{5.14}$$

From (5.5) and (5.10) it is easy to see that

$$d(\omega \vee \eta) = d\omega \vee \eta + (-1)^r \omega \vee d\eta \tag{5.15}$$

where $\eta$ is any $s$-form (and $\omega$ is an $r$-form) in $\mathcal{B}_2$. From the rules of exterior multiplication and (5.5), it is easy to see that

$$\eta \vee \omega = (-1)^{rs} \omega \vee \eta. \tag{5.16}$$

From (5.4), (5.12), (5.14), and (5.16), one derives

$$(* \omega, * \eta) = (\omega, \eta) \tag{5.17}$$

If $M$, $\omega$, and $\zeta$ are all smooth and $\omega$ and $\zeta$ are of the same kind and degrees $r$ and $r-1$, respectively, we obtain

$$(\delta\omega, \zeta) = (-1)^{1+n(r-1)} (* d * \omega, \zeta) = (-1)^r (d * \omega, * \zeta)$$

$$= (-1)^r \int_M d * \omega \vee ** \zeta = (-1)^{r+(r-1)(n-r+1)} \int_M d * \omega \vee \zeta$$

$$= (-1)^{r+(r-1)(n-r+1)} \int_M [d * \omega \vee \zeta + (-1)^{n-r} * \omega \vee p\zeta$$

$$+ \int_M d\zeta \vee * \omega = (d\zeta, \omega) = (\omega, d\zeta)$$

since the first integral vanishes by Stoke's theorem for $(n-1)$-forms, the bracket being just $d[*\omega \vee \zeta]$ (see (5.15)). We emphasize the result:

$$(\delta\omega, \zeta) = (\omega, d\zeta). \tag{5.18}$$

In the case of smooth manifolds and forms, we see from (5.10) and (5.14) that

$$d(d\omega) = \delta(\delta\omega) = 0. \tag{5.19}$$

Combining this with (5.18), we see that

$$(\delta\alpha, d\beta) = 0. \tag{5.20}$$

The formulas (5.18) and (5.20) can be extended to $\mathfrak{B}_2$ forms on manifolds only of class $C_1^1$ by using a proper partition of unity (recall Lemma 1.3), such that if the supports of two of the $h_i$ intersect then their union lies in one coordinate patch, to represent each form as a sum of forms whose supports have the same property. Then, for instance

$$(\delta\omega, \zeta) = \sum_{r,s} (\delta\omega_r, \zeta_s)$$

and each term may be evaluated using one coordinate patch; in that patch, the $g_{ij}$ and the forms may be approximated by smooth forms.

In the case of a coordinate system of the type in (5.7) where we also assume that all the $\partial g_{ij}/\partial x^k = 0$ at $x_0$, we see from (5.10) and (5.13) that the components of $d\omega$ at $x_0$ are

$$\begin{aligned}(d\omega)_{i_1\ldots i_{r+1}} &= \sum_{q=1}^{r+1} (-1)^{q-1}\, \omega_{i_1\ldots i_{q-1} i_{q+1}\ldots i_{r+1}, i_q}(x_0)\\ (\delta\omega)_{i_1\ldots i_{r-1}} &= (-1)^r \sum_{s=1}^{n-r+1} (-1)^{s-1}\, \omega_{i_1\ldots i_{r-1}\, l_s, l_s}\end{aligned} \tag{5.21}$$

where $i_1 \ldots i_{r-1}\, l_1 \ldots l_{n-r+1}$ is a permutation. From (5.21), we see that Dirichlet integral $D(\omega)$ in (5.11) reduces to

$$D_0(\omega) = \int_G \sum_{(i)\alpha} \omega_{(i),\alpha}^2\, dx + 2 \sum_{(i)(j)} \sum_{\alpha\beta} \int_G [\omega_{(i),\alpha}\, \omega_{(j),\beta} - \omega_{(j),\beta}\, \omega_{(j),\alpha}]\, dx \tag{5.22}$$

for the case that $\omega$ has support in a coordinate patch having domain $G$ and the $g_{ij} \equiv \delta_{ij}$ throughout $G$; the last integrals all vanish in this case.

We now prove the following important lemma, first proved for forms by Gaffney

LEMMA 5.1: *Given $\varepsilon > 0$, $0 \leq r \leq n$, and $P_0$ on $M$, there is an admissibile coordinate system mapping $B(0, \varrho)$, for some $\varrho > 0$, onto a neighborhood $U$ of $P_0$, and a constant $l$ such that*

$$(5.23) \qquad D(\omega) \geq (1-\varepsilon) \int_{B(0,\varrho)} \sum_{(i)\alpha} \omega^2_{(i),\alpha} \, dx - l(\omega, \omega)$$

*for any $r$-form $\in \mathfrak{B}_2$ whose support is in $U$.*

*Proof:* We begin by choosing a fixed coordinate system mapping some $B_R = B(0, R)$ onto a neighborhood $U_R$ of $P_0$, carryng the origin into $P_0$, and satisfyng $g_{ij}(0) = \delta_{ij}$. From our formulas for $d\omega$ and $\delta\omega$, we see that

$$(5.24) \qquad D(\omega) = \int_{B_\varrho} [a^{(i)(j)\alpha\beta} \omega_{(i),\alpha} \omega_{(j),\beta} + 2b^{(i)(j)\alpha} \omega_{(i),\alpha} \omega_{(j)} + c^{(i)(j)} \omega_{(i)} \omega_{(j)}] \, dx$$

where the $a$'s are combination of the $g_{ij}$ only and so are Lipschitz and the $b$'s and $c$'s are combinations of the $g_{ij}$ and their first derivatives and so are bounded and measurable at least. Since the $a$'s are Lipschitz and since

$$|2\alpha\beta| \leq \eta\alpha^2 + \eta^{-1}\beta^2$$

we see that we may choose $\varrho$ so small that

$$D(\omega) \geq \left(1 - \frac{\varepsilon}{2}\right) D_0(\omega) - \frac{\varepsilon}{2} \int_{B_\varrho} \sum_{(i)\alpha} \omega^2_{(i),\alpha} \, dx - l(\omega, \omega)$$

The result follows from (5.22).

The following important theorem corresponds to Gårding's Inequality for differential equations:

THEOREM 5.4: *For each $r = 0, \ldots, n$ and coordinate covering $\mathcal{U}$ of $M$, there exist constants $K_{\mathcal{U}} > 0$ and $L_{\mathcal{U}}$ such that*

$$(5.25) \qquad D(\omega) \geq K_{\mathcal{U}} ((\omega, \omega))_{\mathcal{U}} - L_{\mathcal{U}} (\omega, \omega)$$

*for ever $\omega \in \mathfrak{B}^r_2$.*

*Proof:* From Theorem 5.2 it is sufficient to prove this for some particular $\mathcal{U}$. Let $\mathcal{U} = (U_1, \ldots, U_Q)$ be an open covering of $M$ by coordinate patches such that each $x \in M$ is in some $U_k$ satisfying (5.23) with $\varepsilon = \frac{1}{2}$, say. Let $G_1, \ldots, G_Q$ be the domain in $E^n$ such that $U_k = Q_k(G_k)$ for all $k$. There exists a finite sequence $\Phi_1, \ldots, \Phi_s$ of Lipschitz functions on $M$, each of which has support interior to some $U_q$, and such that

$$\sum_{s=1}^{S} \Phi_s(x) = 1 \tag{5.26}$$

for all $x \in M$.

Now if (5.25) were false for the $\mathcal{U}$ just described, there would exist a sequence $\{\omega_p\}$ of $r$-forms in $\mathcal{B}_2^r$ such that $D(\omega_p)$ and $(\omega_p, \omega_p)$ were uniformly bounded but $\| \omega_p \|_U \to \infty$. Then, for some $s$, $q$, and some subsequence, still called $\omega_p$, we would have

$$\int_{G_q} \sum_{(i),\alpha} (\Phi_s \, \omega_{p(1)}^{(q)})^2_{x^\alpha} \to \infty$$

where $\Phi_s$ has support in $U_{q'}$ since

$$\| \omega_p \|_{\mathcal{U}} \leq \sum_{s=1}^{s} \| \Phi_s \, \omega_p \|_{\mathcal{U}}$$

and

$$\| \Phi_s \, \omega_p \|^2_{\mathcal{U}} = (\Phi_s \, \omega_p, \Phi_s \, \omega_p) + \sum_{s=1}^{Q} \int_{G_q} \sum_{(i),\alpha} (\Phi_s \, \omega_{p(i)}^{(q)})^2_{x^\alpha} \, dx \, .$$

But it is easy to see that $D(\Phi_s \omega_p)$ and $(\Phi_s \omega_p, \Phi_s \omega_p)$ are uniformly bounded. From our choice of neighborhoods we have reached a contradiction with the fact that

$$D(\Phi_s \, \omega_p) \geq \frac{1}{2} \int_{G_q} \sum_{(i),\alpha} (\Phi_s \, \omega_{p(i)}^{(q)})^2_{x^\alpha} \, dx \, .$$

We can now present the variational method. We begin with the following lemma:

**LEMMA 5.2:** *Let $\mathcal{M}$ be any closed linear manifold in the space $\mathcal{L}_2^r$ of $r$-forms on $M$ (of some one kind). Then either there is no form $\omega$ of $\mathcal{M}$ which is in $\mathcal{B}_2^r$ or there is a form $\omega_0$ in $\mathcal{M}\mathcal{B}_2^r$ with $(\omega_0, \omega_0) = 1$ which minimizes $D(\omega)$ among all such forms.*

*Proof:* If $\mathcal{M}$ contains no form in $\mathcal{B}_2^r$, there is nothing to prove. Otherwise let $\{\omega_k\}$ be a minimizing sequence, i. e., one such that $(\omega_k, \omega_k) = 1$ and $\omega_k \in \mathcal{M} \cap \mathcal{B}_2^r$ for each $k = 1, 2, \ldots$, and such that $D(\omega_k)$ opproaches its infimum for all $\omega \in \mathcal{M} \cap \mathcal{B}_2^r$. From Theorem 5.4 it follows that the

$((\omega_k, \omega_k))_{\mathcal{U}}$ are uniformly bounded. Accordingly, a subsequence, still called $\{\omega_k\}$, exists which converges weakly in $\mathcal{B}_2^r$ to some form $\omega_0$. But from Theorem 5.3 $\omega_k$ tends strongly in $\mathcal{L}_2^r$ to $\omega_0$ and $D(\omega)$ is lower-semicontinuous with respect to weak convergence in $\mathcal{B}_2^r$. The proof of the lemma is now complete.

DEFINITION: A *harmonic field* $\omega$ on $M$ is a form in $\mathcal{B}_2$ on $M$ for which $d\omega = \delta\omega = 0$ almost everywhere. We will let $\mathcal{H}^r$ denote the linear manifold of harmonic fields on $M$ of degree $r$. (Strinctly speaking we have $\mathcal{H}_i^r$ and $\mathcal{H}_0^r$ for even and odd forms, respectively).

THEOREM 5.5: *For each* $r = 0, \ldots, n\ (= \dim M)$ *the linear manifold* $\mathcal{H}^r$ *is finite dimensional.*

*Proof*. The $\mathcal{B}_2$ forms are dense in $\mathcal{L}_2^r$, since the Lipschitz forms are. Let $M_1 = \mathcal{L}_2^r$. There is a form $\omega_1$ in $M_1 \cap \mathcal{B}_2^r$ which minimizes $D(\omega)$ among all such forms with $(\omega, \omega) = 1$. Let $M_2$ be the closed linear manifold in $\mathcal{L}_2^r$ orthogonal to $\omega_1$, and let $\omega_2$ be the corresponding minimizing form in $M_2$. By continuing this process, we may determine successive minimizing forms $\omega_1, \omega_2, \omega_3, \ldots$, each satisfying $(\omega_k, \omega_k) = 1$ and being orthogonal to all the preceding ones.

Now if $D(\omega_1) > 0$, there are no harmonic fields $\neq 0$ since $D(\omega_1) \leqq \leqq D(\omega_2) \leqq \ldots$. On tue other hand, suppose $D(\omega_k) = 0$ for all values of $K$. Then by Theorem 5.4, $((\omega_k, \omega_k))_{\mathcal{U}}$ is uniformly bounded in $k$, whence a subsequence $\{\omega_p\}$ converges weakly in $\mathcal{B}_2^r$ and hence strongly in $\mathcal{L}_2^r$ to some form $\omega_0$ in $\mathcal{B}_2^r$. This is impossible since the $\omega_k$ form an orthonormal system in $\mathcal{L}_2^r$.

THEOREM 5.6: *For each coordinate covering* $\mathcal{U}$ *of* $M$ *there is a constant* $\lambda_0$ *such that*

$$D(\omega) \geqq \lambda_0 ((\omega, \omega))_{\mathcal{U}} \tag{5.27}$$

*for any* $\omega$ *in* $\mathcal{B}_2^r$ *wich is orthogonal to* $\mathcal{H}^r$.

*Proof*. For, let $\omega_0$ be that form in $\mathcal{B}_2^r$ (there is one since each harmonic field is in $\mathcal{B}_2$) which minimizes $D(\omega)$ among all $\omega$ in $\mathcal{B}_2^r$ with $(\omega, \omega) = 1$ and $\omega$ orthogonal to $\mathcal{H}^r$. Then clearly $D(\omega_0) > 0$ and by homogeneity

$$D(\omega) \geq D(\omega_0)(\omega, \omega)$$

for all $\omega$ in $\mathcal{B}_2^r$ and orthogonal to $\mathcal{H}^r$. By Theorem 5.4 we see that

$$K_{\mathcal{U}}((\omega, \omega))_{\mathcal{U}} \leqq \{1 + L_{\mathcal{U}}/D(\omega_0)\} D(\omega),$$

from which (5.27) follows.

THEOREM 5.7: *Suppose $\omega_0$ is any form in $\mathcal{L}_2^r$ and orthogonal to $\mathcal{H}^r$. Then there is a unique form $\Omega_0$ in $\mathcal{B}_2^r$ and orthogonal to $\mathcal{H}^r$ such that*

$$(5.28) \qquad (d\,\Omega_0\,,d\zeta) + (\delta\Omega_0\,,\delta\zeta) = (\omega_0\,,\zeta)$$

*for every $\zeta$ in $\mathcal{B}_2^r$. Moreover, the transformation from $\omega_0$ to $\Omega_0$ is a bounded linear transformation from $\mathcal{L}_2^r$ into $\mathcal{B}_2^r$.*

*Proof:* From Theorem 5.5, we see that

$$I(\omega) \equiv D(\omega) - 2(\omega\,,\omega_0) \geq \lambda_0 \|\omega\|_{\mathcal{H}}^2 - P\|\omega_0\|_{\mathcal{H}}$$

since $(\omega\,,\omega_0)$ is a bounded linear functional on $\mathcal{B}_2^r$; here $\|\omega\|_{\mathcal{H}}^2 = ((\omega,\omega))_{\mathcal{H}}$. Hence $I(\omega)$ is bounded below and is lower-semicontinuous with respect to weak convergence in $\mathcal{B}_2^r$ if $\omega$ is orthogonal ($\mathcal{L}_2$-sense) to $\mathcal{H}^r$. Accordingly there is a minimizing form $\Omega_0$. If $\zeta$ is any form in $\mathcal{B}_2^r$ orthogonal to $\mathcal{H}^r$, we then see that

$$(5.29) \quad I(\Omega_0 + \lambda\zeta) = I(\Omega_0) + 2\lambda\,[d\Omega_0\,,d\zeta) + (\delta\Omega_0\,,\delta\zeta) - (\omega_0\,,\zeta)] + \lambda^2 D(\zeta)$$

which shows that (5.28) holds for all such $\zeta$ and $\Omega_0$ is unique. But then (5.28) holds all $\zeta$ in $\mathcal{B}_2^r$ since any such $\zeta$ is uniquely representable in the form $\mathcal{G} = H + \zeta_0$ where $dH = \delta H = 0$ and $\zeta_0$ is in $\mathcal{B}_2^r$ and orthogonal to $\mathcal{H}^r$. Finally, if we set $\zeta = \Omega_0$ in (5.28) and use Theorem 5.7, we see that

$$\|\Omega_0\|_{\mathcal{H}}^l \leq \lambda_0^{-1} \|\Omega_0\|_{\mathcal{H}} \cdot \|\omega_0\|_{\mathcal{H}}$$

from which the last statement follows.

DEFINITION: The form $\Omega_0$ of Theorem 5.7 is called the potential of $\omega_0$.

We observe that if all forms in (5.28) and the manifold $M$ were sufficiently smooth, the equation (5.28), together with equation (5.18) would imply that

$$(5.30) \qquad \Delta\,\Omega_0 \equiv d\,\delta\Omega_0 + \delta\,d\Omega_0 = \omega_0\,.$$

In any coordinate system, (5.30) reduces to a system of second order equations in the components of the forms; if $r \geq 1$, these equations involve the second derivatives of the $g_{ij}$ as well as those of the components of $\Omega_0$. However, all the results stated so far hold for manifolds of class $C_1^1$ in which case the requisite second derivatives of the $g_{ij}$ certainly do not exist.

**Definition:** We say that $\omega$ *is of class* $\mathcal{L}_{2\lambda}$, $0 \leq \lambda < n/2$, if for each coordinate system $\theta$ with domain $B_R$, there is a constant $L = L(\theta, \omega)$ such that

$$\int_{B_r} \bar{\omega}_{(i)}^2 \, dx \leq L^2 r^{2\lambda}, \; 0 \leq r \leq R \; (B_r = B(0, r)).$$

The class $\mathcal{B}_{2\lambda}$ is defined similarly.

The importance of the spaces $\mathcal{B}_{2\lambda}$ ariees from the fact that if $\omega \in \mathcal{B}_{2\lambda}$ with $\lambda = \mu - 1 + n/2$, $0 < \mu < 1$, then $\omega \in C_\mu^0$; this follows from the strainghtforwand extension of Lemma 4.1, to $n$ dimensions. We can now state the following results concerning differentiability.

**Theorem 5.8:** *Suppose that* $\omega \in \mathcal{L}_2^r \ominus \mathcal{H}^r$ *and* $\Omega$ *is its potential.*

(i) *If* $M$ *is of class* $C_1^1$, *the* $\Omega$, $d\Omega$, *and* $\delta\Omega \in \mathcal{B}_2$.

(ii) *If* $M$ *is of class* $C_1^1$, *and* $\omega \in \mathcal{L}_{2\lambda}$, *then* $\Omega$, $d\Omega$, *and* $\delta\Omega \in \mathcal{B}_{2\lambda}$ *and hence in* $C_\mu^0$ *if* $\lambda = n/2 - 1 + \mu$, $0 < \mu < 1$.

(iii) *If* $M$ *is of class* $C_1^1$ *and* $\omega \in \mathcal{B}_2$, *then* $d\Omega$ *and* $\delta\Omega$ *are the potentials of* $d\omega$ *and* $\delta\omega$, *respectively.*

(iv) *If* $M$ *is of class* $C_\mu^k$ *and* $\bar{\omega} \in C_\mu^{k-2}$, $k \geq 2$, $0 < \mu < 1$, *then* $\Omega$, $d\Omega$ *and* $\delta\Omega \in C_\mu^{k-1}$. *If* $k \geq 3$ *and* $\omega \in C_\mu^{k-3}$, *then* $\Omega \in C_\mu^{k-1}$.

(v) *If* $M$ *and* $\omega$ *are of class* $C^\infty$ *or analytic, then so is* $\Omega$. *In all case, if we set* $\alpha = d\Omega$ *and* $\beta = \delta\Omega$ *we have*

$$(5.31) \qquad \delta\alpha + d\beta = \delta(d\Omega) + d(\delta\Omega) = \omega, \; d\alpha = \delta\beta = 0.$$

**Theorem 5.9:** *Suppose that* $H$ *is a harmonic field.*

(i) *If* $M \in C_1^1$, then $H \in \mathcal{B}_{2\lambda}$ with $\lambda = n/2 - 1 + \mu$ for any $\mu$, $0<\mu<1$.

(ii) *If* $M \in C_\mu^k$, $k \geq 2$, $0 < \mu < 1$, *then* $H \in C_\mu^{k-1}$.

(iii) *If* $M \in C^\infty$ *or is analytic, then so is* $H$.

In both Theorems 5.8 and 5.9, 0-forms have an additional degree of differentiability (except in the second part of Theorem 5.8 (iv)). It should be observed that we can form $\Delta\,\Omega$ as indicated in (5.31) even though the individual components of $\Omega$ do not have the necessary second derivatives (if $r > 0$).

*Proof:* Obviously $H$ satisfies (5.28) with $\omega_0 = 0$. Then equations (5.28) are a special case of the more general equations

$$(5.32) \qquad (d\omega - \varphi, d\zeta) + (\delta\omega - \psi, \delta\zeta) = (\omega_0, \zeta)$$

Using (5.24) and (5.22) we see that equations (5.32) are equivalent to equations of the form (4.13), if $\zeta$ has support on some one coordinate patch, where the $a$'s are Liptschitz, the $b$'s and $c$'s are bounded and measurable, and the $e's$ and $f's \in \mathcal{L}_2$. Such systems have been studied extensively by the writer in [75] and [47]. Since Professor Nirenberg's lectures are concerned with differentiability problems, the results and their proofs are omitted.

The results concerning $\Omega$ and $H$ follow directly from the result just mentioned. To prove the differentiability of $d\Omega$ and $\delta\Omega$, we select a coordinate patch and find that we can approximate to $\Omega$, $\omega$, and the $g_{ij}$ by smoot functions so That $\Omega$ is a potential of $\omega$ with respect to the altered $g_{ij}$ at each stage. Then, if $\zeta$ has support interior to this patch, we see that (5.31), (5.18), and (5.20) imply that $\alpha$ and $\beta$ satisfy

$$(5.33)\qquad \begin{aligned} &(d\alpha, d\zeta) + (\delta\alpha - \omega, \delta\zeta) = 0 \\ &(d\beta - \omega, d\zeta) + (\delta\beta, \delta\zeta) = 0 . \end{aligned}$$

The interior boundedness theorem (like Theorem 4.5) and an approximation theorem for such systems allow us to pass to the limit in (5.33). If $\omega \in \mathcal{B}_2$, we use (5.33) and (5.18) to see that $\alpha$ and $\beta$ are the potentials of $d\omega$ and $\delta\omega$, respectively.

The following theorem complements the well-known orthogonal decomposition of Kodaira [29].

THEOREM 5.10: *If $\omega$ is any form in $\mathcal{L}_2$, then there exists a harmonic field $H$ and forms $\alpha$, $\beta$, and $\Omega$ in $\mathcal{B}_2$ such that*

$$(5.34)\qquad \begin{aligned} &\omega = H + \delta\alpha + d\beta, \; d\alpha = \delta\beta = 0, \\ &\alpha = d\Omega, \; \beta = \delta\Omega, \end{aligned}$$

*where $\Omega$ is the potential of $\omega - H$. If the first equation of (5.34) holds for a harmonic field $H_1$ and forms $\alpha_1$ and $\beta_1$ in $\mathcal{B}_2$, then $H_1 = H$, $\delta\alpha_1 = \delta\alpha$, and $d\beta_1 = d\beta$.*

*The sets $\mathcal{C}^r$ or all forms $\delta\alpha$ for $\alpha$ in $\mathcal{B}_2^{r+1}$ and $\mathcal{D}^r$ of all forms $d\beta$ for $\beta$ in $\mathcal{B}_2^{r-1}$ are closed linear manifolds in $\mathcal{L}_2^r$ and*

$$(5.35)\qquad \mathcal{L}_2^r = \mathcal{H}^r \oplus \mathcal{C}^r \oplus \mathcal{D}^r .$$

*If $M \in C_1^1$ and $\omega \in \mathcal{L}_{2\lambda}$ or $\mathcal{B}_{2\lambda}$, $0 \leq \lambda < n/2$, then $\delta\alpha$ and $d\beta$ have the same properties.*

*If $M \in C_\mu^k$ and $\omega \in C_\sigma^l$ with $k \geq 2$, $0 < \mu < 1$, $0 < \sigma < 1$, and either $l < k - 1$ or $l = k - 1$ and $\sigma \leq \mu$, then $\delta\alpha$ and $d\beta$ have the same differentiability properties as $\omega$.*

*If $M$ and $\omega \in C^\infty$ or are analitic, so are $\delta\alpha$ and $d\beta$.*

*Proof:* The first statement and the differentiability results follow immediately from Theorems 5.8 and 5.9 If $H, \alpha$, and $\beta$ all $\in \mathcal{B}_2$ (and have properly related degrees), formulas (5.18) and (5.20) and the definition of harmonic field imply that $H, \delta\alpha$, and $d\beta$ are orthogonal in $\mathcal{L}_2$. To see that the sets $\mathcal{C}^r$ and $\mathcal{D}^r$ are closed we see, by following the construction in the first paragraph of the theorem with $\omega = \delta\alpha$ and $d\beta$ in turn, that if $\alpha$ and $\beta \in \mathcal{B}_2$, there are forms $\alpha_1$ and $\beta_1$ in $\mathcal{B}_2$ and orthogonal to $\mathcal{H}$ such that

$$\delta\alpha_1 = \delta\alpha, \ d\alpha_1 = 0, \ \delta\beta_1 = 0, \ d\beta_1 = d\beta.$$

Then if $\delta\alpha_n \to \sigma$ in $\mathcal{L}_2$, we see that the $\alpha_{1n} \to$ some $\alpha_1$ in $\mathcal{B}_2$ by Theorem 5.6. A corresponding result holds if $d\beta_n \to \tau$ in $\mathcal{L}_2$.

## BIBLIOGRAPHY


[1] N. ARONSZAJN and K. T. SMITH, *Functional spaces and functional completion,* Ann. Inst. Fourier Grenoble 6 (1956), 125-185.

[2] F. E. BROWDER, *Strongly elliptic systems of differential equations,* Contributions to the theory of partial differential equations, 15-51. Ann. of Math. Studies, No 33. Princeton University Press (1954).

[3] — , *Numerous notes in Proc. Nat. Acad. Sci.* U.S.A. 38 (1952), 230-235 and 741-747 ; 39 (1953), 179-184 and 185-190 ; and many others.

[4] J. W. CALKIN, *Functions of several variables and absolute continuity,* I, Duke Math. J. 6 (1940), 170-185.

[5] L. CESARI, *An existence theorem of the calculus of variations for integrals on parametric surfaces,* Amer. J. Math. 74 (1952); 265-295.

[6] S. CINQUINI, *Sopra l'estremo assoluto degli integrali doppi in forma ordinaria,* Ann. Math. Pura Appl. (4) 30 (1949) 249-260.

[7] R. COURANT, *Plateau's problem and Dirichlet's principle,* Ann. of Math. (2) 38 (1937) 679-724.

[8] — , *Dirichlet's principle, conformal mapping, and minimal surfaces,* Interscience Press, New York (1950).

[9] J. M. DANSKIN, *On the existence of minimizing surfaces in parametric double integral problems in the calculus of variations,* Rev. Math. Univ. Parma 3 (1952)· 43-63.

[10] E. DE GIORGI, *Sull'analiticità delle extremali degli integrali multipli,* Atti Accad. Naz. dei Lincei Rend. Cl. Sci. Fis. Mat. Nat. (8) 20 (1956), 438-441.

[11] J. DENY, *Les potentiels d'énergie fini,* Acta Math. 82 (1950), 107-183.

[12] G. DE RHAM and K. KODAIRA, *Harmonic Integrals* (mimeographed notes), Institute for Advanced Study, Princeton, N. J. (1950).

[13] G. F. D. DUFF and D. C. SPENCER, *Harmonic tensors on Riemannian manifolds with boundary,* Ann. of Math. 56 (1952), 128-156.

[14] G. C. EVANS, *Fundamental points of potential theory,* Rice Inst. Pamphlets 7 (1920), 252-359.

[15] — , *Note on a theorem of Bôcher,* Amer. J. Math. 50 (1928), 123-126.

[16] — , *Potentials of positive mass I.* Trans. Amer. Math. Soc. 37 (1935), 226-253.

[17] G. FICHERA, *Esistenza del minimo in un classico problema en calcul delle variazioni,* Atti Accad. Naz. Lincei Rend. Cl. Sci. Fis. Mat. Naz. (8) 11 (1951), 34-39.

[18] K. O. FRIEDRICH, *On the identity of weak and strong extensions of differential operators,* Trans. Amer. Math. Soc. 55 (1944), 132-151.

[19] — , *On the differentiability of the solutions of linear elliptic differential equations,* Comm. Pure Appl. Math. 6 (1953), 299-326.

[20] — , *On differential forms on Riemannian manifolds,* Comm. Pure Appl. Math. 8 (1955), 551-558.

[21] G. FUBINI, *Il principio di minimo e i teoremi di esistenza per i problemi di contorno relativi alle equazioni alle derivate parziali di ordine pari,* Rend. Circ. Mat. Palermo 23 (1907), 58-84.

[22] M. P. CAFFNEY, *The harmonic operator for exterior differential forms*, Proc. Nat. Acad. Sci. U.S.A. 37 (1951), 48-50.

[23] —, *The heat equation method of Milgram and Rosenbloom for open Riemanniam manifolds*, Ann. Of Math. 60 (1954), 458-466.

[24] L. GARDING, *Dirichlet's problem for linear elliptic partial differential equations*, Math. Scad. 1 (1953), 55-72.

[25] W. V. D. HODGE, *A Dirichlet problem for harmonic functionals with applications to analytic varieties*, Proc. London Math. Soc. (2) 36 (1935), 257-303.

[26] —, *The Theory and Applications of Harmonic Integrals*, Second Edition, Cambridge University Press 1952.

[27] E. HOPF, *Zum analytischen Charakter der Lösungen regulärer zweidimensionaler Variationsprobleme*, Math. Zeit. 30 (1929), 404-413.

[28] F. JOHN, *Derivatives of continuous weak solutious of linear elliptic equations*, Comm. Pure Appl. Math. 6 (1953) 327-335.

[29] K. KODAIRA, *Harmonic fields in Riemannian manifolds*, Ann. of Math. 50 (1949), 587-665.

[30] P. D. LAX, *On Cauchy's problem for hyperbolic equations and the differentiability of the solutions of elliptic equations*, Comm. Pure Appl. Math. 8 (1955), 615-633.

[31] H. LEBESGUE, *Sur le problème de Dirichlet*, Rend. Circ. Mat. Palermo 24 (1907), 371-402.

[32] B. LEVI, *Sul principio di Dirichlet*, Rend. Circ. Mat. Palermo 22 (1906). 293-359.

[33] H. LEWY, *On minimal surfaces with partially free boundary*, Comm. Pure Appl. Math. 4 (1952), 1-13.

[34] L. LICHTENSTEIN, *Uber den analytischen Charakter der Lösungen zweidimensionaler Variatiosprobleme*, Bull. Acad. Sci. Cracovia, Cl. Sci. Mat. Nat. (A) (1912), 915-941.

[35] —, *Zur Theorie der Konforme Abbildung. Konforme Abbildung nichtanalyticher singularitatenfreier Flachenstüche auf ebene Gebiete*, Bull. Int. Acad. Sci. Cracovia, Mat. Nat. Cl. (A) (1916), 192-217.

[36] E. J. MCSHANE, *Integrals over surfaces in parametric form*, Ann. of Math. 34 (1933), 815-838.

[37] —, *Parametrizations of saddle surfaces with application to the problem of Plateau*, Trans. Amer. Math. Soc. 35 (1934), 718-733.

[38] A. MILGRAM and P. ROSENBLOOM, *Harmonic forms and heat conduction*, Proc. Nat. Acad. Sci. U.S.A. 37 (1951), 180-184 and 435-438.

[39] C. B. MORREY, Jr. *On the solutions of quasi-linear elliptic partial differential equations*, Trans. Amer. Math. Soc. 43 (1938), 126-166.

[40] —, *Functions of several variables and absolute continuity II*, Duke Math. J. 6 (1940), 187-215.

[41] —, *A correction to a previous paper*, Duke Math. J. 9 (1942), 120-124.

[42] —, *Multiple integral problems in the calculus of variations and related topics*, Univ. of Calif. Publications in Math. new ser. 1 (1943), 1-130.

[43] —, *The problem of Plateau on a Riemannian manifold*, Ann. of Math. 49 (1948), 807-851.

[44] —, *Quasi-convexity and the lower-semicontinuity of multiple integrals*, Pac. J. Math. 2 (1952), 25-53.

[45] —, *Second order elliptic systems of differential equations*, Ann. of Math. Studies No. 33 Princeton University Press, 1954, 101-159.

[46] —, *A variational method in the theory of harmonic integrals*, II Amer. J. Math. 78 (1956), 137-170.

[47] C. B. MORREY and J. EELLS Jr., *A variational method in the theory of harmonic integrals* I, Ann. of Math. 63 (1956), 91-128.

[48] M. NAGUMO, *Uber die gleichmässige Sumierbarkeit und ihre Anwendung auf ein Variationsproblem,* Jap. J. Math. 6 (1929), 173-182.

[49] J. NASH, *Continuity of solutions of parabolic and elliptic equations,* Amer. J. Math. 80 (1958), 931-954.

[50] O. NIKODYM, *Sur une classe de fonctions considerées dans l'étude du problème du Dirichlet,* Fund. Math. 21 (1933), 129-150.

[51] G. NÖBELING, *Uber die erste Randwertaufgabe bei regularen Variationsproblemen* I. Math. Zeit. 51 (1949), 712-751.

[52] H. RADEMACHER, *Uber partielle und totale Differenzierbarkeit von Funktionen mehrerer Variabeln, und uber die Transformation der Doppelintegrale,* Math. Ann. 79 (1918), 340-359.

[53] F. RELLICH, *Ein Satz Uber mittlere Konvergence,* Gottingen Nach. (Math.-Phys. Kl.) (1930), 30-35.

[54] H. SAKS, *Theory of the Integral,* Sec. Ed., Warsw-Lwow, 1937; English translation by L. C. Young.

[55] — , *On the surfaces without tangent planes,* Ann. of Math. 34 (1933), 114-124.

[56] L. SCHWARTZ, *Theorie des distributions* I *et* II, Actualites Sci. Ind. 1091 et 1122, Publ. Inst. Matc. Univ. Strasbourg 9 et 10, Paris 1950 et 1951.

[57] M. SHIFFMAN, *Differentiability and analyticity of solutions of double integral variational problems,* Ann. of Math. 48 (1947), 274-284.

[58] A. G. SIGALOV, *Conditions for the existence of a minimum of double integrals in un unbounded region,* Doklady Akad. Nauk SSSR. (N.S.) 8 (1951), 741-744 (Russian).

[59] — , *Regular double integrals of the calculus of variations in non-parametric form,* Doklady Akad. Nauk SSSR (N.S.) 73 (1950), 891-894 (Russian).

[60] — , *Two dimensional problems of the calculus of variations in non-parametric form. transformed into parametric form,* Mat. Sbornik NS 94 (76) (1954), 385-406.

[61] — , *On conditions of differentiability and analiticity of solutions of two dimensional problems of variations,* Doklady Akad. Nauk SSSR (NS) 85 (1952), 273-275.

[62] G. I. SILOVA, *Existence of an absolute minimum of multiple integrals in the calculus of variations,* Doklady Akad. Nauk SSSR (N.S.) 102 (1955), 699-702.

[63] — , *Two dimensional problems of the calculus of variations,* Uspehi Matem. Nauk (N.S.) 6 (1951), 16-101 (Russian).

[64] S. SOBOLEV, *Sur quelques evuluations concernant les familles des fonctions ayant des derivees a carré integrable,* Comptes Rend. Acd. Sci. SSSR. N.S.I. 1936, 279-282.

[65] — , *On a theorem of functional analysis,* Mat. Sbornik N.S. 4 (1938). 471-497.

[66] D. C. SPENCER, *Dirichlet's principle on manifolds,* Studies in Mathematics and Mechanics presented to Richard von Mises, Academic Press (1954), 127-134.

[67] G. STAMPACCHIA, *Sopra una classe di funzioni in due variabili. Applicazioni agli integrali doppi del calcolo delle variazioni,* Giorn. Mat. Battaglini (4) 3 (79) (1950), 169-208.

[68] — , *Gli integrali doppi del calcolo delle variazioni in forma ordinaria,* Atti Accad. Naz. Lincei Rend. Cl. Sci. Fis. Mat. Nat. (8) 8 (1950), 21-26.

[69] — , *Sistemi di equazioni di tipo ellittico a derivate parziali del primo ordine e proprietà degli estremi degli integrali multipli,* Ricerche Mat. 1 (1952), 200-226.

[70] — , *Problemi al contorno per equazioni di tipo ellittico e derivate parziali e questioni di calcolo delle variazioni connesse,* Ann. Mat. Pura Appl. (4) 33 (1952), 211-238.

[71] L. Tonelli, *Sulla quadratura delle superficie,* Atti Accad. Naz. Lincei Rend. Cl. Sci. Fis. Mat. Nat. (6) 3 (1926), 633-638.

[72] — , *Sui massimi e minimi assoluti del calcolo della variazione,* Rend. Circ. Mat. Palermo 32 (1911), 297-337.

[73] — , *Sul caso regolare nel calcolo delle variazioni,* Rend. Circ. Mat. Palermo 35 (1913), 49-73.

[74] — , *Sur une méthode directe du calcul des variations,* Rend. Circ. Mat. Palermo 39 (1915), 233-264.

[75] — , *La semicontinuità nel calcolo delle variazioni,* Rend. Circ. Mat. Palermo 44 (1920), 167-249.

[76] — , *Fondamenti del calcolo delle variazioni,* Bologna, Zanichelli, 3 vols.

[77] — , *Sur la semi-continuité des intégrales doubles du calcul des variations,* Acta Math. 53 (1929). 325-346.

[78] — , *L'estremo assoluto degli integrali doppi,* Ann. Sc. Norm. Pisa (2) 3 (1933), 89-130.

[79] L. Van Hove, *Sur l'extension de la condition de Legendre du calcul des variations aux integrals multiples a plusieurs fonctions inconnues,* Nederl. Akad. Wetensch. 50 (1947), 18-23.

GUBBIO - SOC. TIPOGRAFICA «ODERISI» 1960

Estratto dagli *Annali della Scuola Normale Superiore di Pisa*
Serie III. Vol. XIV. Fasc. IV (1960)

# UNIFORMIZZAZIONE E MODULI (*)

di L. BERS (New York)

Le equazioni di Beltrami a coefficienti discontinui furono considerate per la prima volta da Morrey (cfr. [3] anche per i riferimenti bibliografici); esse si sono dimostrate utili nello studio dell'uniformizzazione e dei moduli delle superficie di Riemann. Nella teoria dei moduli che se ne deduce, ci si è limitati naturalmente al « caso classico » ; tuttavia alcuni teoremi possono essere dimostrati anche pel caso di superficie di Riemann aperte.

Una presentazione completa dell'argomento è ovviamente impossibile in poche pagine. Ci limiteremo perciò solo ad enunciare il risultato centrale per le superficie di Riemann chiuse (§ 1), risultato che commenteremo al § 2; quindi al § 3 daremo un teorema di uniformizzazione che è essenziale per la dimostrazione e che è di per sè interessante. Le dimostrazioni saranno date per sommi capi nei § 6, 7, 8. Una esposizione completa sarà pubblicata in seguito.

## § 1. — Enunciato del teorema fondamentale.

Il teorema che enunceremo assicura in sostanza la possibilità dell'uniformizzazione simultanea di tutte le curve algebriche di dato genere $g > 1$. Esso è da ravvicinarsi al teorema corrispondente per le funzioni di Weierstrass $p(z, 1, \tau)$ e $p'(z, 1, \tau)$, $|z| < \infty$, $\Im m\, \tau > 0$ che dà l'uniformizzazione delle curve di genere 1.

---

(*) Lavoro eseguito col contratto No. DA — 30 — 069 — ORd — 2153 dell'« *Office of Ordonance Research* » dell'Esercito degli Stati Uniti. Questo lavoro uscirà in inglese nei Rendiconti del « *Symposium on function theory* » del « *Tata International Institute* ». Il contenuto di questo lavoro è stato esposto in un ciclo di conferenze tenute presso la Scuola Normale Superiore di Pisa dal 1 al 10 settembre 1958 in occasione del corso internazionale organizzato dal CIME e tenuto sotto gli auspici della Scuola Normale Superiore e dell'Istituto Matematico dell'Università di Pisa.

TEOREMA: Sia $g > 1$ un intero fissato. Esiste:

1) un *dominio limitato* $T$ nello spazio numerico complesso $\mathbb{C}^{3g-3}$ (ove $\tau_1, \ldots, \tau_{3g-3}$ sono le coordinate) omeomorfo ad una cella

2) un dominio $M \subset \mathbb{C}^{3g-2}$ (ove $Z, \tau_1, \ldots, \tau_{3g-3}$ sono le coordinate) omeomorfo ad una *cella* e olomorficamente equivalente ad un dominio limitato.

3) una funzione continua $\sigma(t, \tau)$ a valori complessi, $-\infty < t < +\infty$, $\tau \in T$ tale che $\sigma(t, \tau)$ è *olomorfa* in $\tau$ per ogni fissato $t$, $\sigma(t, \tau) \to \infty$ per $\tau$ fissato e $|t| \to \infty$, $\sigma(t_1, \tau) \neq \sigma(t_2, \tau)$ se $t_1 < t_2$.

4) un gruppo $G$ di automorfismi *analitici complessi* di $M$ che opera su $M$ senza punti fissi e in modo *propriamente discontinuo.*

5) un gruppo $\Gamma$ di automorfismi *analitici complessi* di $T$, *propriamente discontinuo* (ma *non* privo di punti fissi).

6) una applicazione *olomorfa* $\tau \to Z(\tau)$ di $T$ nello *spazio di* Siegel delle coppie $Z = X + iY$ di matrici $g \times g$ simmetriche $X$, $Y$ con $Y > 0$.

ed infine

7) un numero finito di funzioni *meromorfe* $F_j(z, \tau)$ definite su $M$ *automorfe* rispetto a $G$.

tali che le seguenti conclusioni siano verificate:

8) per ogni $\tau \in T$ la curva $\gamma(\tau): z = \sigma(t, z)$, $-\infty < t < +\infty$ è la curva frontiera di un *dominio semplicemente connesso* $D(\tau)$ nel piano della variabile $z$.

9) un punto $(z, \tau_1 \ldots \tau_{3g-3}) = (z, \tau)$ è in $M$ se e solo se $\tau \in T$ e $z \in D(\tau)$.

10) ogni elemento di $G$ è della forma

$$\tau \to \tau \; .$$

$$z \to \frac{a(\tau) z + b(\tau)}{c(\tau) z + d(\tau)}$$

ove $a$, $b$, $c$, $d$, sono funzioni olomorfe in $T$ e $ad - bc = 1$. $G$ è generato da $2g$ elementi $A_1, \ldots, A_g, B_1, \ldots, B_g$ tali che $\Pi[Aj, Bj] = 1$ ove $[A, B] = A B A^{-1} B^{-1}$

11) $G(\tau)$, la «restrizione» di $G$ per $\tau$ fissato, è un gruppo di trasformazioni di Möbius del dominio $D(\tau)$ in sè. La superficie di Riemann $S(\tau) = D(\tau)/G(\tau)$ è una *superficie chiusa di genere g.*

12) *Ogni* superficie di Riemann chiusa di genere $g$ è conformemente equivalente ad una $S(\tau)$, $S(\tau')$ e $S(\tau'')$ sono conformemente equivalenti se e solo se $\tau'$ e $\tau''$ sono equivalenti rispetto a $\Gamma$.

13) La matrice $Z(\tau)$ è una *matrice di Riemann di periodi* per $S(\tau)$, còrrispondente ad una base dell'omologia definita da $A_i$, $B_i$,

e finalmente

14) Le restrizioni delle funzioni $F_j$ per $\tau$ fissato generano il corpo delle funzioni automorfe di $D(\tau)$ rispetto a $G(\tau)$ cioè il *corpo delle funzioni meromorfe* su $S(\tau)$.

Osserviamo il seguente

COROLLARIO I. Lo spazio $T/\Gamma$ è uno spazio analitico complesso normale. Questo segue da 5) e da un teorema di H. Cartan [9]. Dimostrazioni essenzialmente diverse sono dovute a Röhrl [12] e Baily [4]. Baily ha anche dimostrato che $T/\Gamma$ è un aperto di Zariski di una varietà algebrica. Lo spazio $T/\Gamma$ è, com'è ovvio, lo spazio delle classi di superficie di Riemann di genere $g$ conformemente equivalenti (cfr. 12)).

COROLLARIO II. Lo spazio $M/G$ è una varietà complessa, l'applicazione naturale $M/G \to T$ è olomorfa, l'immagine inversa di $\tau \in T$ è una sottovarietà regolarmente immersa in $M/G$ e conformemente equivalente a $S(\tau)$.

In modo analogo si possono costruire dei fissati complessi su $T$ per i quali la fibra su $\tau$ è $S(\tau) \times \ldots \times S(\tau)$ ovvero la varietà di Jacobi di $S(\tau)$.

## § 2. — Osservazioni.

Descriveremo in questo § gli elementi necessarii alla dimostrazione.

*A*) Notazioni. La lettera $S$ denoterà una superficie di Riemann astratta. Si dirà che $S$ è eccezionale se $S$ ammette automorfismi non identici conformi omotopi all'automorfismo identico. Una $S$ *non eccezionale* si può rappresentare come $U/G$ ove $U$ è il semipiano di Poincaré e $G$ è un gruppo Fuchsiano privo di trasformazioni ellittiche; diremo che $S$ è di *prima specie* se i punti fissi di $G$ sono densi sull'asse reale.

In particolare sia $S$ *di tipo* $(g, n)$, cioè ottenuta da una superficie di Riemann chiusa di genere $g \geq 0$ sopprimendovi $n \geq 0$ punti distinti. Allora $S$ non è eccezionale (e di prima specie) se $3g - 3 + n > 0$.

Sia $m$ un differenziale di tipo $(-1, 1)$ su $S_0$. Localmente $m = \mu(\xi)\, d\bar{\xi}/d\xi$ ove $\mu$ è una funzione misurabile e $\xi$ una uniformizzazione locale. Poichè $|\mu|$ è uno scalare possiamo definire $\|m\| =$ estremo superiore essenziale di $|\mu|$. Se $\|m\| < 1$ scriveremo $m \in B(S_0)$ e diremo che $m$ è un *differenziale di Beltrami*. In tal caso $S_0^m$ denota la superficie $S_0$ munita della struttura conforme definita dalla condizione: ogni soluzione dell'*equazione di Beltrami* $\xi_{\bar{\xi}} = \mu \xi_{\xi}$ è una funzione olomorfa su $S_0^m$ (Si richiede che la soluzione sia continua ed abbia derivate generalizzate localmente a quadrato integrabile). L'applicazione naturale $S_0^m \to S_0$ si noterà con $l$.

Un omeomorfismo $S \xrightarrow{f} S_0$ si dice *quasi-conforme* se può essere fattorizzato nel modo seguente $S \xrightarrow{h} S_0^m \xrightarrow{l} S_0$, $h$ essendo conforme. Supponiamo $f$ quasiconforme e sia $[f]$ la classe d'omotopia di $f$. Diremo che $(S, [f], S_0)$ è una *coppia pari.* Due tali coppie $(S, [f], S_0)$ ed $(S', [f'], S'_0)$ si diranno equivalenti se esistono applicazioni conformi $S' \xrightarrow{h} S$, $S'_0 \xrightarrow{h'_0} S_0$ tali che $[h_0^{-1} f h] = [f']$; fortemente equivalenti se $S'_0 = S_0$, mentre $h_0$ può essere assunta eguale all'identità. Oggi coppia pari è equivalente ad una del tipo $(S_0^m, [l], S_0)$. Per abuso di linguaggio identificheremo spesso coppie con le corrispondenti classi di equivalenza.

*B*) Sia $S_0$ non eccezionale. L'insieme delle classi di equivalenza forte di coppie pari $(S, [f], S_0)$ è lo *spazio di Teichmüller* $T(S_0)$. La *distanza di Teichmüller* tra $(S, [f], S_0)$ ed $(S', [g], S_0)$ è data da $\log \{(1+k)/(1-k)\}$ ove

$$k = \inf \| m \| \qquad \text{per} \qquad m \in B(S),$$

$(S^m, [l], S)$ fortemente equivalente a $(S', [f^{-1}g], S)$. Questa distanza definisce una topologia su $T(S_0)$.

Una funzione continua $\Phi$ a valori complessi su $T(S_0)$ sarà chiamata *analitica-complessa* o *olomorfa* se, per ogni insieme $(m_1, \ldots, m_r) \subset B(S)$, ove $(S_1 [f], S_0) \in T(S_0)$, la rappresentazione di un intorno di $0 \in C^r$ in $C$ espressa dalla

$$(\xi_1, \ldots, \xi_r) \to p\,(S^{\xi_1 m_1 + \ldots + \xi_r m_r}, [f], S_0) \to \Phi(p)$$

è olomorfa. In modo analogo definiremo l'*analiticità reale.*

Una rappresentazione quasi conforme $S_1 \xrightarrow{g} S_0$ induce una « rappresentazione lecita » $g^*$ di $T(S_1)$ su $T(S_0)$:

$$g^*((S, [f], S_1)) = (S, [gf], S_0);$$

$g^*$ dipende soltanto da $[g]$ e *conserva* la distanza di Teichmüller e l'analiticità reale e complessa. Il gruppo delle rappresentazioni lecite di $T(S_0)$ in sè sarà denotato con $\Gamma(S^0)$.

Ricorrendo all'uniformizzazione mediante gruppi Fuchsiani si dimostra che: $T(S_0)$ è uno *spazio metrico completo*; se il gruppo fondamentale di $S_0$ è generato in modo finito, $\Gamma(S_0)$ è *propriamente discontinuo*; le funzioni analitiche reali su $T(S_0)$ *separano i punti.* In base al nuovo teorema di uniformizzazione enunciato nel § 3 si dimostra che: se $S_0$ è di *prima specie*, le funzioni olomorfe su $T(S_0)$ *separano i punti.*

*c*) Se $S_0$ è di tipo $(g, n)$ scriveremo $T(S_0) = T_{g,n}$ $\Gamma(S_0) = \Gamma_{g,n}$. Questa notazione è giustificata dal fatto che due qualunque superficie di tipo $(g, n)$ sono *quasi-conformemente equivalenti.* Porremo $\varrho = 3g - 3 + n$, ed assumeremo $\varrho > 0$.

La teoria di Teichmüller [1, 5, 13, 14] delle rappresentazioni quasi conformi estremali implica che $T_{g,n}$ *sia una* $2\varrho$*-cella.* Inoltre è noto che $T_{g,n}$ è una *varietà analitica complessa* (ciò è stato dimostrato per la prima volta da Ahlfors [2]; cfr. anche [6, 10, 11, 15]).

Nel nostro teorema fondamentale, $T$, $\Gamma$ e $M$ tengono il luogo, rispettivamente, di $T_{g,0}$, $\Gamma_{g,0}$ e $T_{g,1}$. Le osservazioni precedenti giustificano alcuni dei nostri enunciati. L'esistenza della rappresentazione descritta in (6), (13) segue, ad esempio, dalla formula variazionale di Rauch [11].

*d*) nell'enunciato del nostro teorema, $T = T_g$ non appare come una varietà analitica complessa astratta ma come un dominio limitato. Questo è un caso particolare di un risultato più generale: $T_{g,n}$ è (olomorficamente equivalente ad) *un dominio limitato in* $C^\varrho$. La dimostrazione (indicata sommariamente in [7] è piuttosto complicata. Essa è basata sulla possibilità di uniformizzare ogni superficie di Riemann chiusa mediante gruppi di Schottky, ed involge una analisi geometrica dettagliata dello spazio di Schottky» di cui $T_g$ è il ricoprimento universale. La dimostrazione procede per induzione su $g$ e su $n$; in tale induzione le superficie iperellittiche rivestono un ruolo particolare.

*e*) la rappresentazione di $T_{g,1}$ nella forma $M$, cioè nella forma descritta negli enunciati 8) e 9) e la costruzione del gruppo $G$ avente le proprietà 4), 10), 11) è basata sul teorema di uniformizzazione del § 3.

Supponendo di aver compiuto le tappe precedenti, non é difficile concludere la dimostrazione, cioè costruire le funzioni $F_j$ aventi le proprietà 7), 14). Fissiamo un insieme di generatori $A_j$, $B_j$ di $G$ (cfr. 10)), e definiamo su ogni $S(\tau)$ una base di omologia, che denotiamo con le stesse lettere. Sia $\omega_j$ il differenziale abeliano di prima specie avente periodo $\delta_{jk}$ su $A_k$ (sicchè, fra l'altro, il periodo di $\omega_j$ su $B_k$ è l'elemento $Z_{jk}$ di $Z(\tau)$). Sia $\Omega_{jk}$ il differenziale abeliano di terza specie su $S(\tau)$ avente periodi 0 sugli $A_j$ e tale che in ogni punto di $S(\tau)$ il residuo di $\Omega_{jk}$ eguagli l'ordine di $\omega_j/\omega_k$. L'insieme delle funzioni $\{\omega_j/\omega_k, \Omega_{jk}/\omega_l, j, k, e = 1, 2, \ldots, g\}$, considerate come funzioni $\mathcal{D}(\tau)$ ha le proprietà richieste.

Osservazione. Il teorema del § 1 è sfortunatamente di carattere piuttosto «esistenziale». Sarebbe utile avere espressioni esplicite per i dominî e le funzioni descritte. Io esito ad affermare che vi sia molta speranza di ottenere tali formule.

## § 3. — Un nuovo teorema di uniformizzazione.

Un gruppo $G$ di trasformazioni di Möbius sarà chiamato *quasi Fuchsiano* se esiste sulla sfera di Riemann una curva di Jordan orientata $\gamma_G$ tale che $\gamma_G$ sia invariante rispetto a $G$, e quest'ultimo sia privo di punti fissi e propriamente discontinuo nei domini $I(\gamma_G)$ e $E(\gamma_G)$ rispettivamente interno ed esterno a $\gamma_G$.

**Teorema I.** Siano $S_1$ e $S_2$ due superficie di Riemann. Supponiamo che $S_1$ e $S_2$ abbiano superficie di ricoprimento universale iperboliche, e che $S_1$ sia *quasi conformemente* equivalente all'immagine speculare $\overline{S}_2$ di $S_2$. In queste ipotesi, esiste un gruppo quasi-fuchsiano $G$ tale che $I(\gamma_G)/G$ sia *conformemente equivalente* a $S_1$ e $E(\gamma_G)/G$ a $S_2$.

Osservazione. $\overline{S}$ è definita sostituendo ciascuna uniformizzazione locale $\xi$ su $S$ con la sua complessa coniugata $\overline{\xi}$. Le ipotesi per il teorema 1 sono soddisfatte se $S_1$ e $S_2$ sono chiuse e dello stesso genere $> 1$.

Dimostrazione. Poniamo $S_0 = \overline{S}_2$. Ne segue che $S_1 = S_0^m$ per un opportuno $m \in B(S_0)$. Per ipotesi $S_0 = \mathcal{U}/G_0$, ove $\mathcal{U}$ è il semipiano superiore e $G_0$ è un gruppo fuchsiano privo di punti uniti. Pertanto $L/G_0 = \overline{S}_0$, $L$ essendo il semipiano inferiore. Poniamo $\mu(z) \equiv 0$ per $\Im z \leq 0$, e definiamo $\mu(z)$ per $\Im z > 0$ mediante la condizione : $\mu(z) d\overline{z}/dz = m$. Risulta $|\mu(z)| \leq k < 1$, e

$$\mu(A(z)) = \mu(z)\overline{A'(z)} \qquad \text{per} \qquad A \in G_0 .$$

Esiste uno ed uno solo omeomorfismo $\omega_\mu$ del piano in sè che lascia 0 e 1 invarianti ed è $\mu$-conforme, ossia è una soluzione dell'equazione di Beltrami.

$$\frac{\partial \omega}{\partial \overline{z}} = \mu(z) \frac{\partial \omega}{\partial z} .$$

Se $A \in G_0$, l'equazione funzionale per $\mu$ implica che $\omega^\mu(A(z))$ è un automorfismo $\mu$-conforme della sfera di Riemann, di guisa che

$$A^\mu = \omega^\mu A (\omega^\mu)^{-1}$$

è una trasformazione di Möbius. Si verifica che $G = \omega^\mu G_0 (\omega^\mu)^{-1}$ è il gruppo quasi-fuchsiano richiesto.

Indicheremo con $\iota$ la rappresentazione naturale di $S_0$ su $\overline{S}_0$. Un omeomorfismo $S \xrightarrow{h} \overline{S}_0$ è detto anti-quasiconforme se può essere fattorizzato nel modo seguente : $S \xrightarrow{h} S_0^m \xrightarrow{\iota} \overline{S}_0$, $h$ essendo conforme. Un $f$ siffatto definisce una *coppia dispari* $(S, [f], \overline{S}_0)$. Ogni coppia dispari è equivalente, nel senso del § 2, *a*), ad una coppia della forma $(S_0^m, [\iota], S_0)$.

Sia $G$ un gruppo quasi-fuchsiano, e poniamo $\Lambda = \Lambda_1/\Lambda_0$, ove $\Lambda_1$ è il gruppo degli automorfismi di $G$ e $\Lambda_0$ è il sottogruppo degli automorfismi interni. Poichè $G$ può essere identificato ai gruppi fondamentali di $I(\gamma_G)/G$ e di $E(\gamma_G)/G$, ogni omeomorfismo

$$I(\gamma_G)/G \xrightarrow{\Phi} E(\gamma_G)/G$$

induce un elemento $\lambda(\Phi)$ di $\Lambda$, e $\lambda(\Phi) = \lambda(\psi)$ se, e soltanto se, $[\Phi] = [\psi]$. Se $\lambda(\Phi) = 1$ e se $\Phi$ è anti-quasiconforme, diremo che $G$ *rappresenta* la coppia $(I(\gamma_G)/G, [\Phi], E(\gamma_G)/G)$ ed ogni coppia equivalente a quest'ultima.

Riesaminando la dimostrazione del Teorema I si vede che di fatto abbiamo dimostrato la prima parte del

TEOREMA II. Ogni coppia dispari $(S, [f], \overline{S}_0)$ può essere *rappresentata* da un gruppo quasi fuchsiano $G$, purchè $S_0$ abbia una superficie di ricoprimento universale iperbolica. Se $S_0$ è di *prima specie, ogni* gruppo quasi-fuchsiano $G_1$ rappresentante questa coppia è della forma $G_1 = QGQ^{-1}$, $Q$ essendo una trasformazione di Möbius.

Per dimostrare la seconda parte dell'enunciato possiamo supporre che sia $S = S_0^m$, $f = \iota$, e che $G$ sia il gruppo costruito dianzi. L'ipotesi fatta su $G_1$ implica l'esistenza di omeomorfismi conformi

$$I(\gamma_G) \cup \gamma_G \xrightarrow{\Phi} I(\gamma_{G_1}) \cup \gamma_{G_1}, \quad E(\gamma_G) \cup \gamma_G \xrightarrow{\psi} E(\gamma_{G_1}) \cup \gamma_{G_1}$$

con

$$\Phi A \Phi^{-1} = \psi A \psi^{-1} \qquad \text{su } \gamma_G \qquad \text{su } A \in G.$$

L'ultima relazione implica che $\Phi = \psi$ in tutti i punti fissi di $G$ e quindi, per continuità, in ogni punto di $\gamma_G$. Poniamo $Q(z) = \Phi(z)$ per $z \in I(\gamma_G) \cup \gamma_G$, $Q(z) = \psi(z)$ per $z \in E(\gamma_G)$. $Q$ è un automorfismo del piano e $G_1 = QGQ^{-1}$.

Resta da dimostrare che $Q(z)$ è olomorfo. Questo sarebbe immediato se $\gamma_G$ fosse rettificabile. Nelle attuali circostanze, tuttavia, dobbiamo considerare $Q(z)$ come funzione di $\xi = \omega^\mu(z)$ ed utilizzare le proprietà di $\omega^\mu$ date, ad esempio, in [3]. Da queste proprietà si trae che mis $\gamma_G = 0$ e che $Q(z)$ ha derivate generalizzate in $L_2$ in un intorno di $\gamma_G$. Poichè $\dfrac{\partial Q}{\partial \bar{\xi}} = 0$ fuori di $\gamma_G$, la analicità di $Q$ ovunque segue da ben note considerazioni.

Siamo ora in grado di costruire il dominio $M$ descritto nel § 1. Nella dimostrazione del Teorema I, sia $S_0$ una superficie prefissata, chiusa e di genere $g > 1$. Supponiamo che per ogni

$$(S_0^m, [f], S_0) = \tau \in T_g = T(S_0),$$

$\omega^\mu$, $A^\mu$ e $G$ abbiano lo stesso significato di prima. Se scegliamo $S_0$ in guisa che $0, 1, \infty$ siano punti fissi, la parte del Teorema II relativa all'unicità mostra che $A$ e $G$ dipendono soltanto da $\tau$ e non dalla scelta particolare di $m$. Possiamo porre $G = G(\tau)$, $\gamma_G = \gamma(\tau)$, $I(\gamma_G) = D(\tau)$. Vale la 11), e $\sigma(t,\tau) = \omega^\mu(t)$. Il fatto che $A^\mu$ e $\sigma$ dipendano olomorficamente da $\tau$ segue da un risultato provato in [3]: se $\mu$ dipende olomorficamente da parametri complessi, altrettanto accade per $\omega^{\mu}(z)$.

Concludiamo enunciando due problemi: Ogni gruppo quasi-funchsiano rappresenta una coppia? È possibile dimostrare il Teorema I «classicamente», cioè usando soltanto trasformazioni conformi?

## BIBLIOGRAFIA


1. L. V. Ahlfors, Journ. d'Analyse Math., 3 (1945), pp. 1-58.
2. L. V. Ahlfors, Analytic functions, Princeton Univ. Press (1959), 45-66.
3. L. V. Ahlfors e L. Bers, In corso di stampa.
4. W. L. Baily. In corso di stampa.
5. L. Bers, *Analytic functions*, Princeton Univ. Press, (1959), pp. 89-119.
6. L. Bers, Proc. Int. Congr. Math. Edinburg 1958, pp. 309-361.
7. L. Bers, Bull. Amer. Math. Soc., 66 (1960), pp. 98-103.
8. L. Bers, ibid., pp. 94-97.
9. H. Cartan, Lefschetz Volume, Princeton Univ. Press (1957), pp. 90-108.
10. K. Kodaira e D. C. Spencer, Ann. of Math., 70 (1959), pp. 145-166.
11. R. H. Rauch, Proc. Nat. Acad. Sci. U. S. A., 41 (1955), pp. 42-49.
12. R. Röhrl, In corso di stampa.
13. O. Teichmüller, Preussische Akad., 22 (1940).
14. O. Teichmüller, Ibid. 4 (1943).
15. A. Weil, Seminaire Bourbaki 1958 (No. 168).


TIPOGRAFIA "ODERISI" EDITRICE - GUBBIO 1960

**Lire 1.800.—**